Survival Strategies

Survival Strategies
Cooperation and Conflict
in Animal Societies

Raghavendra Gadagkar

HARVARD UNIVERSITY PRESS
Cambridge, Massachusetts
London, England
1997

LIBRARY OF CONGRESS CATALOGING-IN-PUBLICATION DATA

Gadagkar, Raghavendra.
  Survival strategies : cooperation and conflict in animal societies /
Raghavendra Gadagkar.
    p. cm.
  Includes bibliographical references (p.   ) and index.
  ISBN 0-674-17055-5 (alk. paper : hardcover)
  1. Social behavior in animals. 2. Animal societies. I. Title.
QL775.G32 1997
591.5—DC21   97-9447

For Vikram

Contents

Preface

"Nothing in biology makes sense, except in the light of evolution." This assertion by evolutionary biologists has only recently percolated to ethology, the study of animal behavior. But the result has been spectacular. Many aspects of the behavior of animals, especially social animals, that appeared paradoxical or could be described but not fully understood, now appear to have a logic. For the first time we are truly able to ask why an animal does what it does. I have attempted to convey this excitement here in simple language, as free of jargon as possible. I have tried to address ordinary people from all walks of life; curiosity about nature is the only prerequisite for reading and, I hope, for understanding what I have written. As is the common practice in ethology, I have freely used words like "selfishness," "altruism," and "nepotism," in talking about the behavior of DNA, chromosomes, cells, and animals. These words carry no moral connotation, but are, as I explain at greater length in Chapter 6, objectively defined in terms of their consequences for the actors.

I was first a Homi Bhabha Fellow and then a B. P. Pal National Environment Fellow while writing this book, and am happy to record my appreciation to J. J. Bhabha, Professor S. M. Chitre, H. D. Pajnigar, and other officials of the Homi Bhabha Fellowships Council for the former

ix

Fellowship and the Ministry of Environment and Forests, Government of India, for the latter Fellowship. My research has been generously supported by the Department of Science and Technology, Ministry of Environment and Forests, Department of Biotechnology, Council for Scientific and Industrial Research, Indian National Science Academy, Jawaharlal Nehru Centre for Advanced Scientific Research, and the Indian Institute of Science, and I thank the authorities of all these organizations. I gratefully acknowledge financial assistance from the Jawaharlal Nehru Centre for Advanced Scientific Research for preparation of the manuscript and the illustrations.

Sanjay Biswas, A. P. Gore, Barbara König, N. Mukunda, and Klaus Riedle made a number of useful and encouraging comments on the manuscript, and I thank them all for their time and their kindness. I thank my wife, Geetha, and my son Vikram for playing the role of natural selection while this book was evolving at the dinner table. Both of them have also read the manuscript and helped clarify my writing more than I care to admit. I thank Swarnalatha Chandran for assistance in preparing the manuscript and L. Geetha for assistance in the library. It was my good fortune that Prema Iyer, Milind Kolatkar, A. V. Narayana, Sanjeeva Nayaka, Sudha Premnath, A. Sumana, and Harry William enthusiastically prepared the illustrations and that E. Hanumantha Rao and S. Sridhar have permitted me to use their excellent photographs. Permission to reprint illustrations from published sources is acknowledged in the figure captions and table sources. I thank *Current Science* for permission to use (with minor modifications) some passages from articles previously published in that journal and H. S. Arathi and Arun Venkataraman for permission to use (with minor modifications) some passages from previously published articles that they co-wrote with me.

Survival Strategies

An Indian tiger on the prowl. *(Photo: E. Hanumantha Rao.)*

1

What Are Social Animals?

The Tiger and the Lion

The tiger and the lion are two of the most majestic predators in the world. But they are also a study in contrast. The tiger, still distributed widely in the forests of Asia, and the subject of massive conservation efforts in India through Project Tiger, is a solitary animal. Apart from a brief association between male and female during courtship and mating and the inevitable association between mother and cubs, the tiger lives and hunts alone. A specialist in stalking and ambushing prey, usually from behind, the tiger is well adapted to living in dense jungle. Once tigers are grown and able to fend for themselves, both males and females leave their birthplaces, perhaps never to return. The male does not even help his mate obtain food; females are known to hunt by themselves even when pregnant.

By contrast, the lion is a gregarious dweller of plains and savannas. Most lions live in Africa, but a few exist precariously in India, where they are restricted to the Gir forest in Gujarat. Not much is known about the Asiatic lion, but George Schaller and others have uncovered intimate details of the lives of African lions. Lions live in prides consisting of

An Indian lion and lioness resting. *(Photo: E. Hanumantha Rao.)*

several adult females (who may be sisters or cousins), subadult males and females, immature cubs, and one or more adult males (who may or may not be related to each other). The members of a pride exhibit a remarkable combination of cooperation and conflict. Their success in hunting comes from coordinating their efforts so that several members of the pride simultaneously attack the prey from different directions. The kill is then shared, but not necessarily in a peaceful, equitable fashion. The males, who usually contribute the least to the hunt, use their muscle power to gain first access to the meat. The females eat next, usually in the order of their social status in the pride. Subadults and cubs

Young Indian lionesses at a kill. *(Photo: E. Hanumantha Rao.)*

have to be content with leftovers and may occasionally die of starvation. The conflict evident in the process of food sharing should not distract us from seeing the contrast between the tiger's solitary habit and the lion's social bent. Cooperation and conflict are inseparable components of any social group of animals, as we will see repeatedly. The fundamental dichotomy is between the tendency to live in groups and the tendency to avoid interactions with other members of the same species. It is hard to say which is the cause and which is the effect, but the solitary habit of the tiger, its preference for cover, and its mode of hunting all reinforce each other. Similarly, the lion's social habit, its preference for the open savannas, and its cooperative hunting strategy seem like un-mistakable components of a different master plan for survival. It is as if the tiger and the lion chose two opposing evolutionary paths in their effort to eke out a living in the harsh wilderness.

The Mosquito and the Bee

The contrast between solitary and social life is not restricted to large animals. Consider the insects. Perhaps the two commonest insects that pierce our skin with a bite or a sting are the mosquito and the bee. The mosquito is as good an example of a solitary creature as the honey bee is of a social one. The female mosquito lays her eggs in stagnant water. The larvae feed solitarily, on microorganisms or if they are carnivorous on other small insects, including the larvae of other mosquito species, and eventually pupate in the water. The emerging adults also lead an entirely solitary life and do not interact with each other, except of course during the brief act of mating. The males feed on nectar and the females usually take at least one blood meal before they lay eggs. For most species, the blood need not come from humans, although some species insist on it. I used to have a colleague who kept his hand inside a cage full of his experimental mosquitoes every day and let them drink his blood. Somehow his commitment to research impressed me even more than that of those whose research involves tracking rogue elephants.

The honey bee, by contrast, cannot live solitarily; bees that lose the way to their hives die in a few hours. One of the greatest biologists of the twentieth century, Karl von Frisch, devoted his life to the study of bees. He discovered color vision in bees, deciphered the bee dance language, and laid the foundations of sensory physiology and of the experimental study of animal behavior. In the words of J. L. Gould, "His pioneering work inspired the discovery of several otherwise unimaginable sensory systems in animals: infrared detectors in night-hunting snakes, ultra-sonic sonar in dolphins and bats, infrasonic hearing in birds, and mag-netic field sensitivity in a variety of animals. Doubtless, other systems are still to be discovered. The lesson is a melancholy one: We are blind to our own blindness, and must not try to read our own disabilities into the rest of the animal kingdom."

There are five species of honey bees—the Afro-European *Apis mellif-era,* and the four Asian species: the domesticated *Apis cerana,* the giant rock bees *Apis dorsata* of the plains and *Apis laboriosa* of the Himalayas, and the dwarf bee *Apis florea.* All species live in large nests made of sheets of wax with hexagonal cells, used both for rearing brood and for storing food. Honey bee larvae are fed on pollen and nectar (adult bees consume only nectar), which the workers collect laboriously. A colony of honey bees may consist of tens of thousands of bees, but only one of them is a queen. All the other female bees are workers, who are much smaller than the queen and also have other morphological adaptations that fit them for their lives as foragers. Depending on the season, the colony may also consist of a small number (a few hundred or less) of males, also called drones. Since the drones do not work for the colony and the queen is virtually an egg-laying machine, all the tasks of nest building, brood care, nest defense, and foraging fall to the workers. Apart from laying a few unfertilized eggs (which develop into drones) in the unlikely event of the queen's death, the workers have no reproductive options of their own. They thus spend their whole lives working and caring for the queen's brood—an act of supreme sacrifice, or altruism.

If that's not sacrifice enough, consider this. The sting of the worker bee is armed with barbs pointing away from its tip so that when firmly lodged in the victim's skin, it cannot be withdrawn. When the bee attempts to fly away after stinging, the sting, the poison gland, and a part of her digestive system are torn away and left attached to the victim.

Apis florea, the Asian dwarf honey bee. These bees are the most primitive of all honey bees, but they exhibit levels of social organization and dance communication similar to those of other honey bee species. Further study of *Apis florea* may well reveal important information about the evolution of honey bees. *(Photo: R. Gadagkar.)*

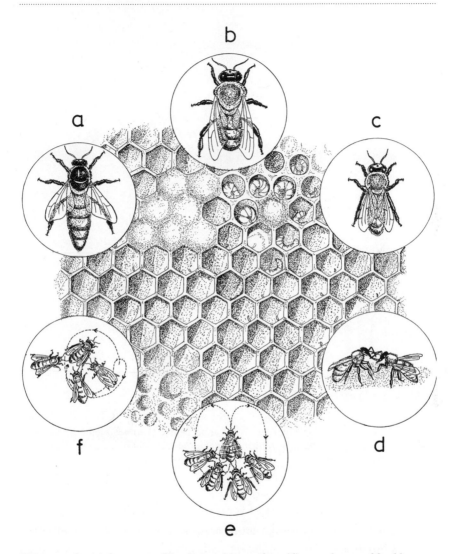

Honey comb: *(a)* the queen, *(b)* a drone, *(c)* a worker, *(d)* an exchange of food between workers, *(e)* the waggle dance, and *(f)* the round dance. *(Drawing: Sanjeeva Nayaka.)*

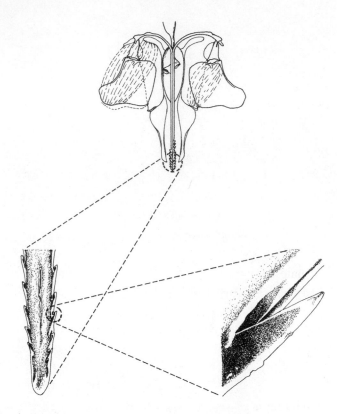

The honey bee worker sting apparatus, with (at bottom left) an enlargement (×360) of the sting and (at bottom right) a further enlargement (×3600) of a single barb. *(Drawing: Sanjeeva Nayaka.)*

This ensures efficient delivery of venom into the victim's body since the poison gland keeps pumping venom for some 30 to 60 seconds after the bee has flown away. But for the bee, stinging is an act of suicide in an attempt to protect the colony. That does not mean there is no conflict in a bee hive; we will see plenty of it later.

Survival of the Fittest?

The worker bee's altruism is not just remarkable from a human point of view; it is also paradoxical from the point of view of Darwin's theory of natural selection. How such altruism evolves in the context of the survival of the fittest is a major puzzle and I will return to it later. Just now we are contrasting the life of the mosquito with that of the honey bee just as we did the life of the tiger with that of the lion. Notice that the pattern of the male lion contributing little to the hunting effort but getting first access to and probably the lion's share of the kill is remarkably similar to the behavior of the drones. The drones stay in the hive of their birth and occasionally in other hives until they are successful at mating, after which they die. But until then they may make several attempts to mate by regularly flying to traditional mating sites where virgin queens from nearby colonies congregate. All along, the drones, who never collect nectar or pollen or help in their storage or distribution to the brood, are fed by the workers. (It is true though that when the food reserves of the colony fall to dangerously low levels, the workers seize the drones by their legs and throw them out of the colony.)

This book is about social animals rather than solitary ones, about the lions and the honey bees rather than about the tigers and the mosquitoes. When animals live in groups, there is a much greater opportunity for complex interactions and hence a greater opportunity for us to see evolution at work shaping animal behavior. Moreover, as human beings, we have a particular curiosity about how animals solve their problems of cooperation, conflict, selfishness, leadership, division of labor, communication, and so on—problems that plague human societies. We need not feel compelled to draw any lessons for ourselves from the knowledge gained about animal social life. Let us merely yield to the pleasures of curiosity. As we have done in the case of the tiger and the lion and the mosquito and the honey bee, we will repeatedly look at examples from diverse animal groups, not only for variety but also to remind ourselves that taxonomic barriers, such as those between vertebrates and invertebrates, birds and mammals, primates and humans, or, indeed, between bacteria and humans, must be broken to see the unity of life in all its glory.

Before we plunge into more details of the social life of animals and possible explanations of why social animals do what they do, we need to define the rules of the game. We will often ask: "Why does an animal behave in the way it does?" First let us see what exactly we mean by such a question and what constitutes an acceptable answer.

The Supreme Sacrifice by Soil Amoebae

The cellular slime mold *Dictyostelium discoideum* normally lives in the soil as individual, free-living microscopic amoebae. The amoebae move about freely, feeding on soil bacteria until the local supply of food is exhausted. Then the best strategy for them is to disperse to a new, food-rich habitat. But the amoebae are simply incapable of this task individually. They therefore go through a social phase to achieve this objective. The hitherto free-living amoebae come together by using an elaborate means of chemical communication and form a multicellular slug. This differentiates into a tall stalk of dead cells on the top of which sits a sphere bearing live spores. Because of the sacrifice of the cells that died to make the stalk, the cells that make up the spores are often able to disperse to a new habitat, where they may germinate to yield new free-living amoebae. And then the cycle repeats itself.

In much the same way as the sterile worker honey bees, the stalk cells have made a supreme sacrifice to enable the spore cells to escape starvation. As in the case of the honey bees, we can ask: "How has such

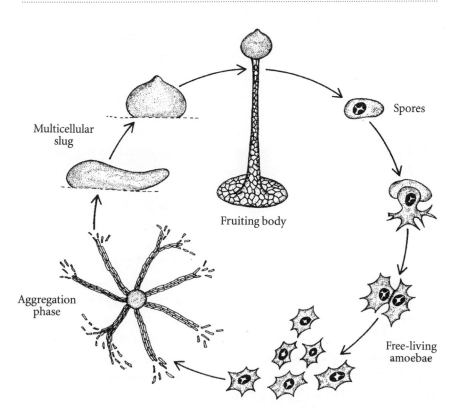

Spores

Multicellular
slug

Fruiting body

Aggregation
phase

Free-living
amoebae

The life cycle of the cellular slime mold *Dictyostelium discoideum.* *(After Olive 1970
and a drawing by Harry William; used by permission of the Indian Academy of Sciences,
Bangalore.)*

altruism on the part of the stalk cells been favored by natural selection?"
I will attempt an answer later, for my intention in bringing up this
example now is to illustrate a special difficulty in asking questions about
animal behavior. A reasonable question is, "Why do the amoebae
aggregate under conditions of starvation?" Physiologically minded re-

searchers have shown that a series of biochemical changes resulting from starvation leads to the production of, and a tendency to move toward, a chemical messenger called cyclic AMP. The amoebae aggregate under conditions of starvation because then they produce cyclic AMP and are attracted to cyclic AMP produced by other amoebae. This is a legitimate answer to the question "Why do the cells aggregate under conditions of starvation?" On the other hand, evolutionarily minded biologists have reasoned that under conditions of starvation, cells aggregate so that at least some of them have a chance of dispersing to a better habitat and surviving. If they did not aggregate, all of them would perish. Mathematical models have been built to show that the spores sitting in the advantageous position on top of the stalk have a good chance of being carried, by wind or by insects, to distant places.

More precisely, the evolutionary explanation goes something like this. The varieties of amoebae that did not have the ability (whatever the nature of that ability) to aggregate and disperse would have died sooner or later. The variety of amoebae that did have the capacity to aggregate and send out at least some of the cells to better habitats thus came to dominate the soil. The varieties that did not aggregate and disperse were *disfavored* while the varieties that aggregated and dispersed were *favored* by natural selection. Now this is an equally legitimate answer to our original question of why the cells aggregate under conditions of starvation.

The Spectacular Migration of the Siberian Cranes

Let's consider another example from a very distant taxonomic group. Every year, Siberian cranes migrate 6400 kilometers from their breeding grounds in Siberia to Bharatpur sanctuary in the state of Rajasthan in northwestern India, where they spend the winter. They arrive in De-

Siberian cranes in Bharatpur, India. *(Photo: E. Hanumantha Rao.)*

cember and leave in March. Our knowledge of the routes of bird migration comes from the ringing of large numbers of birds, from chance sightings by observers along the migration routes, and from the painstaking collation of the resulting information on sightings and ringings. Very little such work has gone on in India, but what little has been done is largely due to the efforts of the late Sálim Ali, the foremost Indian ornithologist, who made birds the best studied vertebrates on the Indian subcontinent. Sought after by the maharajas of colonial India to do bird surveys in their kingdoms, Sálim Ali boasted of having tasted tiger meat and yet did more for the conservation of India's wildlife then anyone else. He laid the foundations of a major study of bird migration at the

Bombay Natural History Society. It is the results of this and many similar studies in different parts of the world that permit us to begin to ask why birds migrate over such spectacular distances and at such precise times of the year. The physiological answer has been worked out to meticulous detail for some species. Shortening day lengths in the northern latitudes are sensed by the pineal gland and lead to hormonal changes that produce "migratory restlessness" and the "urge" to migrate. The evolutionary explanation relates to the net ecological advantages of wintering in the warmer, southern latitudes, in spite of the cost of migrating, compared with the chances of surviving the winter in the northern latitudes.

One Question, Two Answers

Human nature being what it is, the physiologist sometimes tends to see his answer as the correct one and the evolutionary answer as an unnecessary elaboration. Conversely, the evolutionary biologist sometimes tends to see his answer as the more fundamental and philosophically correct one and feels that the physiologist, in his insatiable thirst for detail, has really missed the main point. These attitudes have caused much unnecessary confusion and debate in the past. Clearly, both answers are correct. It's just that they are answers at two different levels. In some situations it seems appropriate to label the proximate explanation an answer to the "how" question and the ultimate explanation an answer to the "why" question. In the slime mold example, the proximate explanation of the biochemical events triggered by starvation appears to answer the question of "how" the slime molds aggregate. But in the Siberian crane example, the proximate explanation of changes in day length leading to altered pineal function and thence to migratory restlessness is not really an answer to the question of "how" the cranes migrate. "How the cranes migrate" begs an answer in terms of how they find their way and how they know where to go. The proximate explanation given above is really an answer to the question of why they migrate in the first place. To label one the answer to the "how" question and the other the answer to the "why" question does not really solve the problem. It is therefore useful to realize that the "why" question in animal behavior can be legitimately answered at at least two levels—the physiological, or *proximate,* level and the evolutionary, or *ultimate,* level. In the long run a complete explanation for animal behavior will require both proximate and ultimate answers, and better still, an integration of the two levels of explanation. But in most cases the time is hardly ripe for that. Without disparaging the proximate explanations in any way I will therefore focus in this book on ultimate, evolutionary answers to a variety of questions about why animals, especially social animals, do what they do. But before I do that, I will take a closer look at evolution by natural selection and define the evolutionary approach to answering questions about animal behavior.

2

Evolution, the Eternal Tinkerer

Pollution and Evolution

Charles Darwin traveled as the ship's naturalist on H.M.S. *Beagle* from December 27, 1831, to October 2, 1836. This gave him an opportunity to observe a tremendous variety of animals and plants and also "to read for amusement Malthus's *Essay on the Principle of Population* (1798)," all of which led over 20 years later to the publication of *The Origin of Species by Means of Natural Selection, or The Preservation of Favoured Races in the Struggle for Life* (1859)—a book that sold out its first printing in one day. It is hard to imagine another book that has so completely changed our view of the world and of ourselves. And yet, perhaps the most remarkable thing about this book is that Darwin never really demonstrates even a single case of evolution in action. Darwin argues convincingly that evolution must have happened, but he never saw it happen. Indeed, he writes that "natural selection is daily and hourly scrutinizing, throughout the world, the slightest variations; rejecting those that are bad, preserving and adding up all that are good; silently and insensibly working, whenever and wherever opportunity offers. . . . We see nothing of these slow changes in progress, until the

16

hand of time has marked the lapse of ages, and then so imperfect is our view into long-past geological ages, that we see only that the forms of life are now different from what they formerly were." For this reason, the moth *Biston betularia* will always remain a textbook example of Darwin's theory of evolution by natural selection in action. Bernard Kettlewell, an English physician who gave up the practice of medicine and turned to the study of *Biston betularia* and other lepidopterans, writes that "among all living things it has fallen to the Lepidoptera to provide evidence of the most striking evolutionary change in nature ever to be witnessed by man."

In the early part of the nineteenth century, the common form of *Biston betularia,* called *typica,* had a peppered appearance. Its wings were flecked with black and white, and it was well camouflaged in its favorite resting place, the pale and lichen-covered barks of trees in rural England. A dark, or melanistic, form of the moth, called *carbonaria,* was first recorded in about 1848, and presumably had existed in very small numbers before then. But by the middle of the twentieth century, the melanistic form of the moth had come to represent over 95 percent of the *Biston betularia* population, especially in such industrial centers as Manchester and Liverpool. This is among the most rapid of all recorded evolutionary changes. Why was there such a dramatic change in fortunes of the peppered *typica* and the melanistic *carbonaria?* With rapid industrialization, soot came to cover the barks of trees, making them black instead of pale, and also killed the lichens. Now the melanistic form was better camouflaged on the darkened bark, while the peppered form became increasingly less camouflaged and hence more easily detected and eaten by birds. The birds, which had kept the melanistic form at a very low frequency before industrialization, now concentrated their attention on the peppered form. Natural selection, in the form of bird predation, favored the peppered form earlier and the melanistic form later.

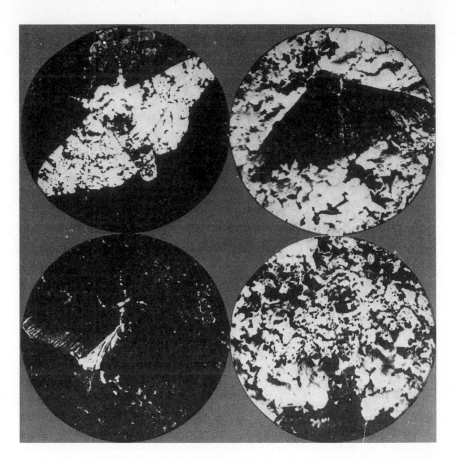

Peppered and melanistic forms of *Biston betularia* resting on lichen-covered and soot-covered backgrounds. Notice the effective camouflage of the peppered form on the lichen background (bottom right) and of the melanistic form on the sooty background (bottom left), and the conspicuousness of the peppered form on the sooty background (top left) and of the melanistic form on the lichen background (top right). *(Reproduced by permission of Oxford University Press from B. Kettlewell, The Evolution of Melanism, 1973.)*

This explanation for the evolution of the melanistic form of *Biston betularia* has been verified in many different ways and found to be reasonably correct. With a few additional minor details, we could recreate the observed changes on a computer. Notice that this means that we are on the right track, because the computer would not reproduce the observed pattern if we input the wrong facts. Of course the ultimate proof would come from demonstrating (in real life, not merely on a computer) that if the environment reverts to its original pristine, unpolluted condition, if the soot disappears and the lichens grow back, the peppered form will regain its dominant position and the melanistic form will gradually disappear. Fortunately, such an ultimate proof has actually been obtained. Strict antipollution laws were adopted in England in 1956 and the countryside became relatively free of smoke. In a mere 20 years, the frequency of the melanistic form dropped significantly, indeed to the levels predicted by the computer models.

More recent research has brought to light strikingly parallel changes in and around Detroit, Michigan, including the near absence of the melanistic form before industrialization, its rise to about 90 percent of the population by 1959–62, and its virtual disappearance by 1994–95, barely 30 years after the clean-air law was adopted in 1963. These parallel and independent changes in England and the United States are like two replications of an experiment with identical results, boosting confidence in our explanation of this phenomenon, which has come to be known as industrial melanism. It must be cautioned, however, that some discrepancies between theoretical expectations and observations remain, suggesting that perhaps we do not know the full story yet; birds that prey on the adult moths appear not to be the only agents of natural selection at work.

The Blind Watchmaker

It is not difficult to see that some win and some lose in the game of Survival of the Fittest in the struggle for existence. Who wins and who loses is determined by the environment, which alone decides who is fitter. But chance often determines which players in the game are present at any given time. If you are not present at the right time, you don't win even if you are fitter. The melanistic *carbonaria* was present, albeit at a very low frequency in the population, when pollution changed the color of the tree bark and thus the fortunes of the two forms of moths. Natural selection did not produce the melanistic form. It must have arisen by chance, and although it was at a disadvantage in the unpolluted environment, natural selection had been unable to kill it off completely or perhaps enough time had not elapsed since its origin for natural selection to have completed the job.

How then did the *carbonaria* form arise? We know that *carbonaria* is a mutant form of *typica*, differing from the latter at just one gene. Both because of toxic chemicals in the cellular environment and inherent errors in the process of duplication, some changes creep into the DNA molecule—the repository of hereditary information. The cellular machinery has an elaborate mechanism to edit the newly made DNA to correct errors. But even so some errors remain. These are called mutations, and they are the raw material of natural selection. *Carbonaria* thus differs from *typica* in carrying just one error. It follows then that natural selection has no purpose, design, or goal; it merely acts on errors that have been lucky enough to escape elimination. And all that natural selection does is to eliminate some of these errors and retain others. It may also blindly change direction and begin to favor the hitherto "unfit" varieties and to kill off the hitherto "fit" ones, when the environment changes in the opposite direction.

How such a blind process of tinkering can produce the most immaculately designed living organisms is mind-boggling to most of us; for some, it is sufficient to abandon the theory of natural selection altogether. At first glance we may find it improbable that natural selection acting on chance mutations could produce complex entities such as the human eye or kidney but we must realize that natural selection had an enormous amount of time available to it for shaping the final products we see today. More important, natural selection does not begin from scratch every time; there is successive selection at each step. There is a famous saying that, given enough time, a monkey typing randomly on a typewriter can produce all the works of Shakespeare. Richard Dawkins, the author of the famous book *The Selfish Gene*, first employing his daughter and later a computer program in place of the monkey, actually tried this experiment, beginning with the simple 28-character-long sentence METHINKS IT IS LIKE A WEASEL. He soon realized that if you had to start from scratch each time ("single-step selection") it would take his computer program about a million million million million million years to hit upon the correct sequence of characters by chance alone. This is not difficult to appreciate. There are 26 characters in the English alphabet, and counting the blank spaces required between words as the 27th character, there is a chance of 1/27 of getting any letter right by chance alone. The probability of getting all the 28 characters right in the required sentence simultaneously would be 1/27 raised to the power of 28, which is equal to about one chance in 10,000 million million million million million million.

But if the computer program were allowed to select, in each generation, the string of 28 characters most closely resembling the target sentence and then to act on further mutants produced from that "best" variety ("cumulative selection"), the job could be done in just about 30 minutes. In one of Dawkins's trials, for example, the computer program began with the phrase WDLMNLT DTJBKWIRZREZLMOQCO P and "mutated" it randomly. Of all the random "mutants" produced, the one most resembling the target sentence was WDLTMNLT DTJBSWIRZREZLMQCO P, and therefore this was chosen as the starting point for the next generation. After 10 generations, the winning phrase was MDLDMNLS ITJISWHRZREZ MECS P, and after 20 generations it was MELDINLS IT ISWPRKE Z WECSEL. After 30 generations the phrase took the form METHINKS IT ISWLIKE B WECSEL, and after 40 generations it became METHINKS IT IS LIKE I WEASEL. At this stage only one letter was incorrect, and it took only 3 more generations to reach the target sentence METHINKS IT IS LIKE A WEASEL.

In opposition to the eighteenth-century theologian William Paley, who argued that just as a watch is too complicated and purposefully designed to have come into existence by accident, living organisms, which are much more complicated, could not possibly have arisen by chance and must have been purposefully designed, Dawkins argues that if natural selection can be said to be a watchmaker, it is a blind watchmaker. Dawkins points out that although the "monkey/Shakespeare model" is useful for explaining the profound difference between single-step selection and cumulative selection, it is misleading in that it gives the impression that each generation is judged by its resemblance to some ideal target. But natural selection has no ideal target, and it worries only about survival in the immediate present. To overcome this misleading implication, Dawkins goes on to use similar computer programs to create animal-like and plantlike shapes by cumulative selection. Having whetted your appetite, I will leave it to you to read about that and indeed to play with such computer programs. There is yet another aspect of Darwin's theory that many people find hard to grasp and that is best illustrated by recounting the shocking behavior of hanuman langurs.

Infanticide among Hanuman Langurs

Presbytis entellus, the hanuman langur, with its black face, gray hair, and long tail, is a spectacular-looking monkey. Its populations range from the Himalayas in northern India to the southernmost tip of the Indian peninsula, extending into Sri Lanka and other land masses on either side of the Indian subcontinent. The epithet "hanuman" comes from the name of the monkey god who helped retrieve Rama's wife, Seetha, from the clutches of Ravana, the king of Lanka, in the Hindu epic *Ramayana*.

Hanuman langurs live either in bisexual troops or in all-male bachelor troops. The bisexual troops consist of several adult females, juveniles of both sexes, and either a single adult male or several adult males. The one-male troop, or harem, is particularly interesting. The male in control of a harem is periodically driven out by an invading bachelor troop. If the invasion is successful, the males of the bachelor troop usually fight among themselves until only one of them retains control of the harem—until he is ousted in a subsequent invasion. Upon taking over a new harem, the male typically kills most or all unweaned infants. How could this behavior, which seems clearly bad for the species, have possibly been favored by natural selection? Not surprisingly, many naturalists have described such infanticide as rare and pathological, and as possibly induced by conditions of overcrowding. Then Sarah Blaffer Hrdy undertook a field study of hanuman langurs in Mt. Abu in Rajasthan, India, from 1971 to 1975, and concluded that infanticide by male langurs taking over new harems was neither pathological nor maladaptive. Even more provocatively, she concluded that it was of great advantage to those males who practiced it and thus could easily have been favored by natural selection. Why this profound difference between Hrdy's attitude toward infanticide and those of previous naturalists?

A hanuman langur mother with her infant. *(Photo: E. Hanumantha Rao.)*

For the Good of the Species?

For over a hundred years following the publication of Darwin's theory, biologists consistently misunderstood an important element of his reasoning, and the "theory of natural selection" they promulgated, which was accepted by both professional biologists and the lay public, was actually a misrepresentation of Darwin's theory of natural selection. Biologists came to substitute for Darwin's precise statements about natural selection their own imprecise version, which may now be called "the good of the species" concept. It became implicit in virtually all discussions of natural selection that evolution works for the good of the species.

The idea that natural selection favors what is good for the species came to an abrupt end in the mid-1960s. The major credit for this complete change in the way we view natural selection must, ironically, go to one of the foremost champions of the idea of the good of the species. This was V. C. Wynne-Edwards, who in 1962 wrote a book, nearly as massive as Darwin's own, in which he attempted to explain a variety of behavior patterns in animals as being designed to promote the good of the species. So far most biologists were interpreting natural selection as promoting the good of the species only in an indirect and vague manner. Wynne-Edwards stuck his neck out and explicitly developed what he believed to be a unified theory of animal behavior and ecology based on the idea that individuals will always be selected to sacrifice their own interest for the sake of the good of the group. It is this clarity and explicitness in Wynne-Edwards's book that immediately made several biologists sit up and realize that there was a major flaw in what had passed for the correct interpretation of Darwin's theory of natural selection.

I once had the good fortune of attending a conference in which the famous evolutionary biologist John Maynard Smith was participating.

The conference was held in the picturesque hill resort of Mahabaleshwar, near Bombay. Perhaps the most vivid impression that I have of that and other Mahabaleshwar conferences I have attended since is the synchronous calls of hundreds of male cicadas, repeated every 30 to 40 minutes. Right in the middle of Maynard Smith's lecture and just when he was describing Wynne-Edwards's theory, there erupted a loud burst of cicada singing. Nonplussed, Maynard Smith said, if Wynne-Edwards were here he would have surely argued that the cicadas are singing in unison so as to assess their population density and adjust the rate of their reproduction so that they do not overexploit the habitat and eventually drive their species to extinction. Indeed, Wynne-Edwards had argued that almost all aspects of animal behavior and ecology were designed to limit their populations so as to avoid destruction of their habitat and their eventual extinction; species that lacked such self-regulating mechanisms would soon go extinct from overexploiting their resources.

The Nobel laureate Konrad Lorenz appears to have fallen into the same trap. He wrote, for example, that "Darwin had already raised the question of the survival of fighting and he has given us an enlightening answer. It is always favorable to the future of the species if the stronger of two rivals takes possession of either the territory or the desired female." Wrongly believing that the killing of conspecifics (other members of the same species) is rare in nature, Lorenz attempted to explain the supposed rarity by arguing that animals either are incapable of killing another of their own kind or must possess "sufficiently reliable inhibitions [to] prevent self-destruction of the species."

Today we know that this reasoning is incorrect and that natural selection is rarely, if ever, concerned with the good of the species. Natural selection almost always acts at the level of individual organisms and selects those that are best adapted to their environment, even if that

A sparring match between male spotted deer *(Axis axis)*. The dominant male will have first access to resources. *(Photo: E. Hanumantha Rao.)*

hurts the group or species as a whole. Most of the natural phenomena that Wynne-Edwards imagined could be explained only by *group selection* are better explained by *individual selection*. Hrdy's individual selection explanation for infanticide in hanuman langurs is that if a male kills unweaned infants immediately after taking over a harem, the females that were hitherto suckling will come to estrus sooner and consequently the male will have higher reproductive success. If he does not kill the infants and waits for them to be naturally weaned, he may sire many fewer offspring. And he often has precious little time before he is ousted by another male. Ideally, he needs to have his own offspring weaned before his ouster so that they are not killed by the next male. Males that practice infanticide under such circumstances will be fitter than those that do not practice infanticide and will increase the representation of their genes in future generations of hanuman langurs. If the propensity to practice infanticide has even a mild genetic component, the noninfanticidal males will eventually disappear and the infanticidal males will come to dominate the population. In many cases we now know that the stronger of two rivals takes possession of the territory or the desired female not because the subordinate male gives up voluntarily for the good of the species, but because accepting the subordinate role is better for him than the risk of injury from a prolonged fight; he will copulate with the desired female as often as possible when the dominant male is not looking (quite unmindful of the good of the species).

One of the arguments made by Wynne-Edwards was that animals will be shaped by natural (group) selection to produce fewer offspring than they can potentially produce so that they do not overexploit their food base. Christopher Perrins studied the swift *Apus apus,* which normally lays two to three eggs but is capable of laying many more, and asked what would happen if more eggs were laid. To answer his question, he artificially increased the number of eggs in some nests to four by adding an extra egg. In each of the four years that he did this, the maximum number of surviving offspring was produced by nests that had three eggs and not by nests that had four eggs. When the parents tried to feed four chicks, they apparently fed each so little food that mortality was higher. Natural (individual) selection thus favors the swifts that lay that number of eggs (three) which results in the largest possible number of surviving offspring, in obvious disregard of the possibility of overexploiting their resource base. Thus the assumption that birds produce fewer offspring than they possibly can is wrong; they seem to lay fewer eggs than they possibly can because the largest possible number of eggs does not lead to the largest possible number of surviving offspring. In other words, they produce as many offspring as they possibly can.

Lorenz's assumption that animals will not kill other members of their species is also wrong. Hrdy writes in the book based on her study of the hanuman langur that "by the time I had concluded my research, I had learned . . . [that] the langur males compete fiercely for possession of females, and that in the process, conspecifics are sometimes killed. Furthermore, langurs are far from unique in this respect. A host of species has been recently added to the list of creatures known to kill conspecifics for motives other than eating them. These include such diverse groups as lions, hippos, bears, wolves, wild dogs, hyenas, rats, rabbits, lemmings, herring gulls, storks, European blackbirds, eagles, and more than fifteen types of primates—or sixteen, counting man."

Citing nobel laureates is an irresistible way of pointing out fallacies that are by no means restricted to their writings; hence the repeated choice of Konrad Lorenz. But it would be wrong to leave the impression that all that Lorenz did was to mix up levels of natural selection. Konrad Lorenz was one of the founders of ethology, the science of the study of animal behavior, and is best loved and remembered for his discovery of imprinting in birds, a discovery he made when birds that he had hand-reared began to treat him as their mother—one bird even tried to court him.

Cheaters Take All

The underlying theoretical reason why Wynne-Edwards's theory will not usually work is that it takes just one cheater to ruin an elaborately laid plan designed for the good of the group. Consider a population of birds in which all members have actually been programmed to produce fewer offspring than they are capable of, so as to ensure sustainable use of their food base. All will be fine until one selfish mutation arises in the population and reproduces as fast as it can. The selfish individuals benefit from the prudence of the altruists and benefit from their survival plan without paying the associated cost of limited reproduction. Eventually the selfish will outnumber the altruists and thus drive the altruists to extinction. In technical parlance, a selfish strategy is stable against invasion by altruists, but an altruist strategy is unstable against and susceptible to invasion by selfish individuals; thus all populations eventually are converted to stable groups of selfish individuals. A cartoon by the famous Gary Larson, showing a band of lemmings on a suicide mission, with one of them wearing an inflated rubber tube around its waist, captures the essential fallacy of Wynne-Edwards's group selection theory better than any verbal description.

It is worth trying to explore the reasons why "the good of the species" idea became nearly universally accepted. There are at least three kinds of reasons that may be adduced. The first is a philosophical one. The idea that individual parts are just slaves in the hands of the master design of nature just seemed more satisfying and correct to people. The idea that the larger unit of organization (the species) ultimately decides the fate of its subcomponents (the individuals) implied a certain harmony in nature. The second reason is a social one. Fairly soon after the publication of Darwin's theory, there began to develop a pseudo-science, often known as Social Darwinism. Proponents of Social Darwinism used their own version of the theory of natural selection to justify human social systems such as capitalism and racism. For example, they argued that there is nothing wrong in the rich getting richer and the poor getting poorer because this is the law of nature—it is natural selection operating for the good of the species. Obviously the good of the species idea came in very handy for such arguments. The third reason is a purely scientific one. Early evolutionary studies concentrated mainly on nonsocial traits, where the good of the individual often coincides with the good of the species. For example, the perfection of the human eye and kidney over evolutionary time is good both for the individuals who possess good eyes and kidneys and for the species as a whole. It was only when social traits, where the good of the species and the good of the individuals do not always coincide, began to be studied, that the fallacy of the concept of the good of the species became clear.

But is it not true that many social animals exhibit altruism, which must benefit their species? Don't honey bee workers sacrifice reproduction and inflict suicidal stings on marauders who invade their nests? Don't stalk cells in the cellular slime mold kill themselves to enable the spore cells to disperse to better habitats? Can all such altruism be explained by individual selection, by the "good of the individual" idea? Could the honey bee worker possibly be selfish and be merely ensuring the propagation of her own genes? The triumph of modern evolutionary biology has been the successful interpretation of nearly all known cases of altruism as a manifestation of some form of individual selection without recourse to Wynne-Edwardian group selection.

Levels of Natural Selection

In describing the triumph of modern evolutionary biology, I deliberately spoke of explaining *nearly* all known cases of altruism on the basis of individual selection because it is not true that natural selection cannot ever act at levels other than the individual. In principle, natural selection can act at any level of biological organization—DNA, genes, cellular organelles, cells, organs, individual organisms, family units, larger groups, populations, species, and even communities of species. Of course the question of where natural selection acts arises only when there is a conflict of interest between different levels of biological organization. If a character (any observable property) simultaneously benefits or harms two levels, then natural selection acts at both levels. For example, if the mitochondria (spherical or rod-shaped bodies inside cells that effect cellular respiration) become more efficient at producing energy, that will simultaneously benefit the mitochondrial genes, the mitochondria itself, the cells and the organs bearing the mitochondria, and the individual organism. The problem arises only when the character in question creates a conflict of interest between different levels of biological organization. If one gene in a cell starts to reproduce faster than necessary for the well-being of the cell and drains the cellular resources, this is good for the gene concerned (at least in the short run) but bad for the cell, the organ, and the individual. Natural selection will then usually act at the level of the cell or individual and suppress such selfish behavior on the part of a gene.

How then do we decide where natural selection will act in a given situation? Can we develop a general theory about this? Perhaps. When two levels of biological organization are competing, as it were, for the attention of natural selection, the strength of natural selection on each level will depend on the relationship between the two levels of organization—how independent the unit at the lower level of organization is of the "clutches" of the higher level of organization, how much short-term gain the units at the lower level can achieve by working against the higher level before they themselves begin to suffer, how much "discipline" the higher level of organization can impose on the lower level. Take for instance a conflict between individual organisms and groups of organisms. The units at the lower level, the individual organisms, are usually pretty free of the clutches of the group; they have a life of their own and can go a long way by revolting against the group. Hence natural selection usually acts at the level of the individual rather than the group. Now consider a conflict between an organism and its constituent cells. The units at the lower level of organization here are the cells. But the cells are pretty much under the control of the higher level of organization, the individual organism. The cells can do precious little to revolt against the whole body and hence natural selection will usually act at the level of the individual organism rather than the cells. We have seen and we will keep seeing examples of natural selection acting at the level of the individual organism. But now let's take up some examples of natural selection acting at other levels, such as the chromosome (DNA), the cell, and groups of organisms.

Selfish DNA

Nasonia vitripennis is a parasitoid wasp that is distributed throughout the world and has a fascinating life cycle. The males have only vestigial wings and cannot fly; they therefore die after mating with females that emerge in the vicinity of their birth. The females, however, can fly, and mated females therefore go off in search of new hosts on which to lay their eggs. Their hosts consist of the pupae of flies that breed in carcasses and in bird nests. Like all insects that belong to the order Hymenoptera, *N. vitripennis* is haplodiploid, meaning that males are haploid, with only one set of chromosomes, while females are diploid, with two sets of chromosomes. The females can lay both fertilized and unfertilized eggs. The unfertilized (and therefore haploid) eggs develop into haploid adult males, while the fertilized (and therefore diploid) eggs develop into diploid adult females. This means that sons have no fathers and fathers have no sons. An equally interesting consequence of this mode of sex determination is that females can decide the sex of their offspring. All they need to do is to release sperm (received at the time of mating and stored in special organs called spermathecae) into their oviducts to produce daughters and block the flow of sperm to produce sons. There is good evidence that females actually utilize this ability to choose the sex of their offspring because they do alter the ratio of haploid to diploid eggs they lay in response to environmental conditions. But there are other genetic and nongenetic factors, not under the control of the females, that can rather drastically alter the sex ratio of the offspring. For example, a bacterial infection can be transmitted from mothers to daughters that kills nearly all haploid eggs, resulting in an all-female line. Such an infected strain cannot survive in the wild unless infected females can find males from other, noninfected families of wasps to keep them going. It is of course easy enough to keep such strains in the laboratory, where one can supply healthy males to each generation of infected daughters.

Another strain of *Nasonia vitripennis* has been found where the opposite happens—only sons are produced. Again it is easy to maintain such a strain in the laboratory by supplying healthy females for every generation of mutant males. John Werren has made headlines with these discoveries, announcing son-killing factors that are passed down from mother to daughter and daughter-killing factors that are passed down from father to son! It is the daughter-killing factor that is of interest here. It is now quite clear that mutant males produce normal sperm with normal-looking chromosomes and that these sperm successfully fertilize eggs. The problem begins after that. For reasons that were not clear earlier, the paternal chromosomes in the sperm disintegrate in the fertilized zygote (the cell formed by the union of two gametes, or sex cells), leaving only the maternal chromosomes. But since the zygote is now haploid it develops into a male rather than into a female. This is how the daughters are killed—or rather converted into sons. How and why do the paternal chromosomes disintegrate? Even more puzzling is the question of how the resulting haploid males get the mutant character so that they will in turn convert potential daughters to sons in the next generation when their sperm fertilizes eggs. Such transmission is not expected because the paternal chromosomes all disintegrate. The only possible conclusion is that the factor that causes the daughter-to-son conversion is extra-chromosomal and comes to the zygote along with the paternal chromosomes and does not disintegrate with them. Careful prying into the structure of the mutant sperm has revealed that mutant males carry a small chromosome (a piece of DNA, if you like) over and above the usual 5 chromosomes that normal males carry. This is called a B chromosome.

In the early days when cytologists were describing chromosomes of various species of plants and animals, they found unusual chromosomes in some species. In addition to the normal sets of chromosomes that are present in pairs in the adult stage and that become haploid in the gametes (sex cells—sperm and eggs) and reunite with another chromosome of their kind during fertilization to restore diploidy, there may occasionally be odd chromosomes that are not usually paired and whose transmission is erratic. They may not be present at all or may be present in variable numbers of copies. Most cytologists did not quite understand the significance of these supernumerary chromosomes and simply labeled them B chromosomes, retaining the label A chromosomes for the apparently normal ones. *Nasonia vitripennis* mutants that show the daughterless phenotype have a B chromosome that reaches the zygote along with the paternal chromosomes and appears to produce a factor that destroys all the paternal chromosomes. But obviously the B chromosome itself is resistant to such destruction, so that it stays on in the resulting haploid cell, which will develop into a male when the B chromosome can do its trick all over again.

The B chromosome confers no benefit to the male that harbored it but instead destroys all the male's chromosomes to ensure its own survival and transmission to future generations. Not surprisingly, the *Nasonia* B chromosome has been dubbed "the most selfish genetic element" known. But obviously this B chromosome can only go so far, because if it invades all males in the population then there will be no females left for the mutant males to mate with. The survival of the B chromosome depends on its ability to use normal females for its onward transmission. Natural selection will therefore restrict the prevalence of the B chromosome to a level low enough that the whole population does not go extinct for lack of females. If the B chromosome does increase in frequency in any population, that population might go extinct, and we can therefore assume that in all surviving populations the B chromosome has been kept under reasonable control. Nevertheless, the *Nasonia vitripennis* B chromosome is an excellent example of how natural selection can sometimes act at levels of biological organization other than the individual organism.

Are Cancer Cells Selfish?

When there is a conflict between cells and the body they reside in, natural selection usually favors the body, which can usually discipline the errant cells, especially because the cells don't have a life of their own outside the body. A well-known exception to this principle is that of cancer cells, which can be thought of as selfish cells attempting to reproduce faster than is good for the health of the whole body. In the end, of course, the cancer cells perish with the individual, but that does not explain why natural selection has not eliminated cancer all together. A common objection to the interpretation of cancer cells as selfish is that they are abnormal and perhaps infected with a virus, that cancer is a disease, and so on. All this is true and pertinent to the proximate answer to the question of why cancer cells reproduce faster than is good for the body. But the ultimate, evolutionary answer must be that natural selection in this case is acting in favor of the cell rather than the individual. The fact that cancer is typically an old-age disease lends further credence to this interpretation; in old age it's no longer critical for the individual to suppress the selfish designs of the cells because the individual has probably already completed its task of reproduction. The very phenomenon of senescence and the prevalence of various other old-age diseases may also be interpreted as resulting from the relaxation of the body's strict control over the selfish tendencies of its organs, tissues, cells, and genes as a person ages.

Altruistic Myxoma Virus in Australia

Australia evolved its own unique mammalian fauna of marsupials and for millions of years did not have the same mammalian fauna as the rest of the world. Rabbits, for example, were unknown in Australia until Europeans introduced them in 1859. But since they did not simultaneously introduce foxes, the rabbits multiplied merrily until they became pests. To control the rabbits, a highly virulent form of the myxoma virus was introduced. This virus was very effective in killing the rabbits, but it went extinct itself whenever the number of rabbits became too small for the virus to travel from one rabbit to another. (The virus depends on mosquitoes to get from one rabbit to another, much like the malarial parasite.) A fresh stock of the virus had to be imported every time the virus became extinct and rabbit populations grew large.

In the course of time a mutation seems to have arisen in the virus population that may be described as an altruistic form. The mutant form of the virus is relatively avirulent and grows rather slowly. This form we may call altruistic because it allows many more virus particles (of its own kind as well as those of other genetically distinct kinds) in the rabbit body to mature before it kills the rabbit. The altruism of course is toward other viruses, not toward rabbits. By contrast, the virulent form of the virus may be described as selfish because it reproduces very fast and uses the resources of the rabbit before the other viruses do so. Here the altruist seems to have defeated the selfish individual. The selfish virulent strain of the virus lost out because it killed the rabbit before the progeny viruses had a chance to be transported by mosquitoes to healthy rabbits. The avirulent viruses kept the rabbit alive for a long time, and consequently mosquitoes efficiently transmitted them from one rabbit to another. Natural selection therefore favored the altruistic avirulent strain over the selfish virulent one.

But such examples are not very common. The simple reason is that the selfish strains usually invade the population and multiply at the expense of other strains. In this particular case, the selfish strain of virus could not easily invade the altruistic population because the selfish viruses killed their host rabbits rapidly and since mosquitoes do not bite dead rabbits they were unable to carry the selfish viruses from one rabbit to another. When there is a conflict between individuals and the group, natural selection usually acts at the level of the individual and promotes selfishness, but the myxoma virus example shows that it can occasionally act at the level of the group and suppress selfishness on the part of the individuals. But this, as we have seen, requires very special conditions indeed.

Before the 1960s, biologists blindly applied the idea of group selection without realizing that natural selection will promote selfishness on the part of individual organisms except under very special circumstances. In mid-1960s and the 1970s, the phrase *group selection* became a term of opprobrium. I have sat in many seminars where a question from a member of the audience was loudly dismissed by other members of the audience shouting "but that's group selection!" even before the speaker had a chance to understand the question. Today the dust has settled down and we recognize that natural selection can, in principle, act at various levels of biological organization and that we must examine the circumstances carefully before pronouncing a judgment about the level of natural selection. This has brought back a level of credibility to mathematical models of group selection that I hope will permit the discovery of more genuine examples of group selection and natural selection at other unexpected levels of biological organization.

3

It's in the Genes

Before we begin to explain the evolution of social behavior, of cooperation and of altruism, by natural selection, acting at the level of the
individual or otherwise, we need to cross one major hurdle. We had
no difficulty in understanding the evolution of the melanistic form of
Biston betularia after industrialization or of the peppered form after
enforcement of antipollution laws, because we knew that the melanistic
and peppered forms were genetically determined. If more melanistic
forms were eaten by birds and more peppered forms were left behind,
we could be certain that the population would come to consist of
more and more peppered forms in future generations. When we argued that the practice of infanticide by hanuman langur males was
beneficial to them and that is why the population has come to consist
of infanticidal males rather than noninfanticidal males, we made the
assumption that infanticidal males are more likely to produce infanticidal sons and noninfanticidal males are more likely to produce noninfanticidal sons. Only under this assumption would it be correct to
argue that natural selection acting on individual male langurs will promote the spread of infanticidal behavior by favoring infanticidal males

and weeding out noninfanticidal males from the population. How good is this assumption?

Hygienic Honey Bees

American foulbrood is a bacterial disease that affects the larvae and kills the pupae of the European honey bee *Apis mellifera*. The bacteria spread through the separating wax wall from infected larvae and pupae to neighboring healthy ones. We don't quite know what natural colonies of honey bees do about it. But beekeepers are usually paranoid about it and immediately burn or bury the affected colonies, wax, honey, workers, queen, and all. Perhaps for this reason, resistance to the disease is not very common. There was, however, one instance when a beekeeper did not destroy diseased colonies. Indeed, it appears that he intentionally acquired diseased colonies, volunteering to dispose of them, but kept them in his apiary so that his healthy bees could steal honey from the diseased colonies. The disease spread to his original colonies too, since the bees stealing honey also stole the disease-causing bacteria from the affected colonies. But this gave the bees a chance to develop resistance to the disease. The resistance they developed was of a most interesting kind. The bees evolved (yes, by random mutation and natural selection) a form of hygienic behavior that let them get one up, or nearly so, on the bacteria. The hygienic worker bees uncap the cells containing dead pupae and remove their corpses.

In the 1960s Walter Rothenbuhler, intrigued by this remarkable behavior of the hygienic bees, decided to investigate it. There were many ways he could have gone about his study, but he took the bold approach of crossing hygienic bees with nonhygienic bees to see what the progeny would do. Now this is not as easy as it sounds. Recall that worker bees don't breed. So one has to cross queens and drones from colonies where workers are hygienic and nonhygienic respectively, or vice versa. This is problematic in the wild, because honey bee queens and drones mate high up in the air and only at specific drone congregation areas where drones and queens from several nearby colonies gather at certain times of the day (and of course only in a particular season). But fortunately beekeepers had developed the technique of artificial insemination for honey bees as early as 1927. So Rothenbuhler's task was not impossible. His results were most intriguing. The first-generation hybrids were all nonhygienic. This means that the forms of the genes (called alleles) controlling nonhygienic behavior are dominant over those controlling hygienic behavior. The standard technique to determine how many genes are involved in specifying any character (hygienic versus nonhygienic behavior, in this case) is to backcross, to cross the hybrids with one of the parents. When Rothenbuhler crossed the first-generation

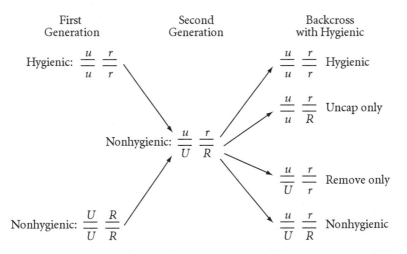

Walter Rothenbuhler's analysis of hygienic behavior in honey bees. U and u indicate the dominant and recessive genes for uncapping behavior, and R and r indicate the dominant and recessive genes for removing behavior.

hybrid with the hygienic parent, he obtained four kinds of queens in equal proportions. One kind produced workers that were hygienic, and the second kind produced workers that were nonhygienic. The third kind produced workers that would uncap the cells but stop at that without removing the dead pupae. The fourth kind produced workers that were equally curious; they would not uncap the cells on their own but would remove dead pupae if Rothenbuhler uncapped the cells for them. The obvious interpretation of these results is that there are two separate genes involved: one controls uncapping of the cells and the other controls removal of the dead pupae.

Hygienic behavior in honey bees was, and perhaps continues to remain, the most complex behavior known to have a very simple genetic basis. Not surprisingly, it has become a standard textbook example of the genetic basis of animal behavior. Having taught this example with great gusto to generations of students, I was crestfallen to read some years ago in an article by Jeffrey Hall that "since [Rothenbuhler's experiments] the entire story has dissolved: The hygienic strain was 'sick' and difficult to maintain." But I was delighted to learn more recently that Marla Spivak of the University of Minnesota has now resurrected the hygienic strain of bees and has initiated work on it, in an attempt to use its hygienic genes to fight a mite infestation that European and American beekeepers appear to have inadvertently imported along with Asian bees.

How Bats and Flies Keep Time

Knowing the time of day is of great survival value to animals. Diurnal animals need to start their activities at dawn and return to safety by dusk. Conversely, nocturnal animals need to start their activities by dusk and return to safety by dawn. Being out of their safe nesting or roosting locations at the wrong time can be dangerous because they may encounter predators to whom they are not accustomed and hence cannot easily escape. Besides, their own prey or other resources, such as nectar and pollen, may be available only at specific times. Sometimes resources are accessible to them only at rather precise times of the day or night, say before the more dominant competitor species come along or after they have left. Having a clock can be crucial for survival. I once

Thousands of Mexican free-tailed bats *(Tadarida brasiliensis)* emerging at dusk from their roosting place in Bracken Cave in Texas. This is much like the scene I witnessed at Madurai. *(Photo: Merlin D. Tuttle, Bat Conservation International.)*

witnessed the unforgettable sight of thousands of bats leaving their cave in Madurai in southern India precisely at dusk, in one big cloud. M. K. Chandrashekaran and his students at Madurai Kamaraj University in the state of Tamil Nadu in India have established an array of research activities in an attempt to understand how the bats know when to leave and when to return.

In one experiment, R. Subbaraj and Chandrashekaran captured some of the bats and kept them individually in laboratory cages. Even in the cages the bats became restless and began to fly about at dusk, like their counterparts in the cave, and quieted down at dawn, when the free bats returned to the cave. Subbaraj and Chandrashekaran then deprived the caged bats of information about when it was dusk and when it was dawn, by keeping them for days under conditions of continuous darkness or continuous light. The bats continued to show very nearly the same periodic bursts of activity every 24 hours or so, but their clock was no longer quite so precise. Sometimes their clock was a bit slow, so that the bats woke up several minutes later each day, and sometimes their clock was a bit fast, so that they woke up several minutes earlier each day. In technical parlance these bats are said to *free run*. Thus some bats had a clock with a periodicity of about 25 hours and others had a clock with a periodicity of about 23 hours. Obviously bats with a clock of 25 hours or 23 hours will soon become out of phase with the outside world. Sure enough, in about 12 days the experimental bats were ready to sleep when the cave bats were just waking up and were waking up when the cave bats were ready to sleep. That is the reason why biological clocks are called *circadian,* meaning "about a day," not *exactly* a day. But why were the clocks precise in the cave or when the laboratory bats were given information about the outside world? As a child, I owned a watch that ran slow by about 10 minutes every day, and after a few days I was hopelessly late for school. I complained bitterly to my father, who did not buy me a new watch but did get me into the habit of resetting my watch every night by listening to All India Radio, and there was no further problem. This is obviously what the bats are doing: resetting (technically called entraining) their clocks daily with reference to the actual times of dusk and dawn. That is why they free run when denied information about the natural day-night cycle of the environment. Clearly this is a superb arrangement since the day length changes from season to season and from place to place. If nature had endowed the bats with a perfect but unresettable clock they would have trouble living in a changing world. They are much better off with an imprecise but resettable clock.

What is the nature of the biological clock and is it specified by genes? Fascinating as they are, bats are not ideally suited to genetic experiments. Perhaps no creature is better suited for genetic experiments than the fruit fly *Drosophila*. In addition to being good for genetic studies, *Drosophila* is also excellent for studies of circadian rhythms. The flies show at least two clear-cut circadian rhythms—the eclosion rhythm (concerning the time of day when they will complete metamorphosis and emerge from the pupal cases), and the locomotor activity rhythm (concerning the time of the day they will be active and the time they will rest). *Drosophila*, which incidentally means lover of dew, ecloses very early in the morning, at about 4:00 A.M. They thus have enough time to stretch and harden their cuticles and begin to fly about before their predators become active. They too have a circadian clock that permits them to do this. Like bats, these flies reset their clocks in accordance with the external world.

Ronald Konopka and Seymour Benzer generated (by means of tricks that we need not worry about here) mutant flies that had abnormal circadian rhythms and showed that a single gene, christened by them *period*, or *per*, is defective in these mutants. The various forms, or alleles, of the *per* locus ("locus" is a fancy word for gene) make the flies have clocks that have cycles (free-run periodicities) of about 16 hours, about 19 hours, or about 28 hours instead of the usual "about" 24 hours; and one form even makes the flies arhythmic. It is remarkable, however, that these short-cycle and long-cycle flies also entrain to the normal 24 hour day-night cycle of the environment and perform the usual rhythmic activities of ordinary flies. It is only when the flies are kept in the laboratory under continuous light or continuous darkness that their mutant nature becomes evident. Incidentally, the defect is seen not only in the eclosion rhythm but also in the locomotor activity rhythm. This engenders some confidence that the *per* gene is close to the heart of the clock and is not just some superficial step in the manifestation of one particular rhythm.

In recent years the *per* gene has been subjected to extraordinarily close scrutiny through the use of the "awesome power of molecular genetics," as one commentator put it. Typically genes (DNA) make a messenger RNA which then makes the protein that performs the required function. The *per* gene protein is made in brain cells but its exact job is not quite clear. What is interesting, however, is that the levels of *per* messenger RNA as well as *per* protein oscillate in the brain cells with a periodicity that mirrors other rhythmic functions. Thus the *per* messenger RNA and protein oscillate with a periodicity of 24 hours in normal and mutant flies entrained to the normal light-dark cycles. The molecular oscillations have a periodicity of *about* 24 hours in normal flies deprived of environmental cues, and a periodicity of the appropriate 16, 19, or 28 hours in the mutants deprived of environmental cues. To complete the story, the messenger RNA and protein do not oscillate in the arhythmic mutant when it is deprived of environmental cues.

I am still not sure whether *per* is the clock itself or whether *per* is only the hand of the clock, and some other master control gene has yet to be discovered. But it is very striking that there appears to be a rather simple genetic basis for such complex behaviors as the timing of eclosion and the locomotor activity cycles of the fruit fly.

Genetically Predisposed Undertaker Bees

As we have already seen, honey bees live in large colonies of several thousand bees. Since each bee lives only about 40 days, there must be a turnover of about a thousand bees every day. Even if a fraction of these dying bees die in the hive, the removal of dead bees represents a major chore for the workers. Not surprisingly, there is a separate caste of worker bees appropriately called undertaker bees. In honey bees different tasks are performed by the same bees at different times in their lives. This age-based division of labor has some leeway for specialization so that some bees are more likely to become undertakers or may remain undertakers longer than other bees. Some others may show a similar preference for guarding behavior, or nursing, or some other task.

What makes some bees more likely to become undertakers? Is it their genes by any chance? Gene Robinson and Robert Page set out to test the idea. Like Rothenbuhler, and incidentally in the bee lab at Ohio State University named after Rothenbuhler, Robinson and Page artificially inseminated queen bees using semen from different drones. They had no way of knowing which drones would produce daughters fond of undertaking; indeed, they did not even know if there was a genetic predisposition for undertaking. Their only criterion for choosing drones as donors of sperm was that the daughters of different drones should be distinguishable from each other biochemically. In colonies containing queens thus artificially inseminated, they captured worker bees that were removing dead bees and bees similar in age that were performing other tasks. Bees fathered by some drones had a higher probability of undertaking than did bees fathered by other drones. Since all the workers were born and brought up in the same colony under the same conditions, the differences could easily be attributable to differences in the genetic make-ups of the undertaker and nonundertaker bees.

At first sight this result seems problematic. Honey bee queens can mate with about 20 drones before settling down to a career of egg laying. But what if a queen fails to mate with a drone that produces undertaker daughters? Won't her colony get clogged by dead bees with no one to remove them? Not at all. The propensity to undertake is not an all-or-none phenomenon; it is mild and relative. Different genetic lines of workers (fathered by different drones who might have inseminated the same queen) have mildly differing tendencies to undertake. This is not surprising. Dead bees stink and different genetic lines of worker bees could easily be imagined to have different levels of sensitivity to the smell of dead bees. Those with the highest sensitivity will be among the first to remove the dead bees. But if a colony does not have any worker particularly sensitive to the smell, the dead bees will continue to decompose and smell even worse. Soon even a bee rather insensitive to the smell will find it intolerable and will do the job.

Today we know that there are similar genetic predispositions for a variety of different tasks, such as nursing, guarding, and grooming. These genetic predispositions for different tasks, coupled with the fact that the queens mate with many males, provide an excellent basis for efficient division of labor in the bee hive. Of concern to us once again is the evidence that genes influence complex behaviors.

The Cricket's Song

The very least anyone knows about crickets is that they sing. (To be precise, crickets cannot sing because they do not vocalize; instead they produce sound by rubbing their hind legs together, a process called stridulation, but for convenience I will continue to use the word "sing.") Typically, males sing a love song to attract females. William Cade has unabashedly pried into the intimate details of cricket romance using the species *Gryllus integer*. The song greatly enhances the male's probability of finding a female. But his singing has a cost associated with it. Because natural selection is dictated by the short-term advantages of chance mutations, anything can evolve so long as it can be produced by mutations and is better than the previous alternative—there are no other rules, no regard for long-term good, and certainly no ethics as we define it.

The cricket love song is exploited in at least two ways. First, some males, often called satellite males, do not sing but listen to the love songs of their rivals and move toward them, much as females do, thus finding females without paying the cost of singing. Locating females is all that counts in the eyes of natural selection; the tactics used are irrelevant.

Saving the energy associated with singing is not the only advantage that the satellite males get. The song is also exploited by a parasitic fly that uses it to locate crickets and deposit its larvae on them. The fly larvae slowly consume the cricket and kill it within the week. The satellite male that cleverly uses its rival's call to locate a female avoids parasites to a large degree.

But every male can't be a satellite. Some have to be singers so that others can live by exploiting them. If the population of singing males goes up, there is scope for more satellites; their strategy becomes more advantageous and their population goes up relative to that of the singing males. At this point the payoffs for the satellites decrease because there are so many of them and there aren't enough singers to attract the females. So the satellites are less successful and their frequency drops.

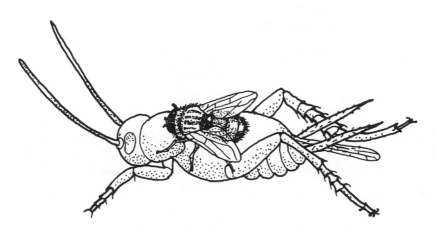

A parasitic fly attracted by the male cricket's song deposits on the cricket live larvae that will eventually kill it. *(Reproduced by permission of Blackwell Scientific Publications from Krebs and Davies 1993.)*

This game may be endless or the singers and satellites may eventually achieve some kind of an equilibrium.

Why are some males shy of singing? Do they assess how many males in the population are singing and then choose their strategy? William Cade artificially selected lines of singers and satellites much as a cattle breeder would select lines to produce high-yielding cows. In each generation he chose individuals with a high propensity for singing for one experiment and individuals with a high propensity for being a satellite in another experiment. In just four generations he obtained lines that differed significantly in singing rates. The satellite line sang only once per night while the singers sang six times per night. Clearly, the complex, alternative strategies of singing to obtain a female and adopting the sneaky satellite strategy of avoiding the cost of singing as well as the cost of being parasitized, are strongly influenced by different genetic constitutions.

The Blackcap's Changing Migration Routes

We have already seen in Chapter 1 that Siberian cranes migrate 6400 kilometers after breeding in Siberia to winter in the milder climate of Bharatpur in Rajasthan. The blackcap *Sylvia atricapilla* is one of the better studied migrants. It breeds in England and northern and central Europe and usually migrates to the warmer Mediterranean regions in the winter. Almost no blackcaps are seen in the winter in England or in northern and central Europe. In the past 30 years or so, the situation has changed dramatically. More and more blackcaps are being seen in England in the winter. The first explanation of this change that comes to mind is that some of the breeders in England simply decided to skip migration and stayed on. But this turned out to be false. Birds ringed in Germany and Austria have been captured in England during the winter, and in recent times almost 10 percent of the birds ringed in Germany and Austria have been recovered in England. The implication is that some of the birds that breed in Germany and Austria have started going to England for the winter rather than to their usual sites in the western Mediterranean.

How do the birds know where to go? A curious fact that makes the experimental tackling of this question possible is that caged birds will demonstrate migratory restlessness in such a way that one can infer when and in what direction they wish to migrate. When kept in cages, blackcaps caught in Germany and Austria orient in a southwesterly direction during their migratory restlessness, consistent with their destination being the western Mediterranean. Birds caught in England in the winter and returned to Germany oriented instead in a northwesterly direction, consistent with England being their destination. When the offspring of birds caught and bred in England in the winter were returned to Germany, they also oriented in a northwesterly direction,

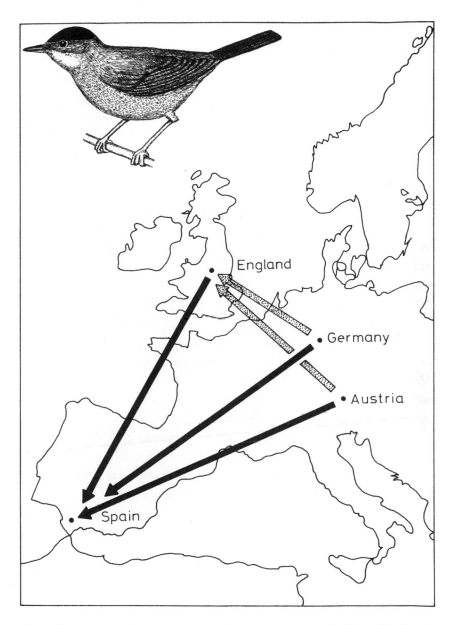

Old (solid arrows) and new (dotted arrows) migration routes of blackcaps *(Sylvia atricapilla)*. *(Drawing: Sanjeeva Nayaka.)*

indicating that even the direction of migration is specified by the genes. As in the case of the moth *Biston betularia,* a fairly rapid evolutionary change appears to have taken place in a short period of time. Recall that the frequency of the melanistic form of the moth rose from a negligible level to about 95 percent in about 100 years. Here the northwesterly migrants increased from a negligible level to about 10 percent in 30 years.

But what is the selective advantage for blackcaps to go to England rather than to the western Mediterranean? Many possibilities exist. The winters in England are becoming milder. The number of people in England who feed birds has grown dramatically. The food these people provide has a significant impact on the birds' survival in winter; indeed, blackcaps and other birds have even begun to gain weight during the English winter. Equally important, England is closer to Germany and Austria than southern Spain and so the migrants have shorter distances to fly. Also they can stay longer in Germany and Austria or return there sooner for the next breeding season. Blackcaps overwintering in England return to their breeding sites almost 10 days earlier than those that come from the western Mediterranean. This head start may be crucial in obtaining territories and beginning the breeding process. Another factor that may accelerate the evolution of northwesterly migration is unconnected to its advantage: If birds from England reach the breeding grounds 10 days ahead of the others, they are likely to mate among themselves and concentrate the genes for northwesterly migration.

Do Genes Determine Behavior?

We have seen that in one way or another genes influence such complex characters as hygienic behavior in honey bees; circadian rhythms in fruit flies; the propensities of honey bees to remove dead bees, guard the nest, and feed and groom other bees; the male cricket's decision to sing or not to sing; and the blackcaps' decision about where to spend the winter. So was our assumption in Chapter 2 that infanticide in hanuman langurs has a genetic basis reasonable? Probably yes, at least for the limited purpose of working out the consequences of natural selection on the infanticidal males.

Does that mean behavior is genetically *determined?* Here our answer should be in the negative. Genes may influence behavior; bearers of certain genes may have higher or lower than average propensities for performing certain behaviors; and parents showing certain behaviors may be more likely to produce offspring that show similar behaviors. But genes seldom, if ever, *determine* behaviors. Behaviors are strongly influenced by the environment, and it would therefore be incorrect to say that they are *determined* by genes, because to do so implies that the environment has no role in the development of behavior. Notice that for natural selection to increase or decrease the frequency of certain behavioral traits, it is not necessary for genes to *determine* behavior. It is sufficient for genes to *influence* behavior. Even if a behavior pattern is the product of interaction between a number of different genes and a number of different environmental factors, natural selection can influence the rate of spread of that behavior. Natural selection may be a bit slow because some offspring of the bearers of a trait may not show that trait as faithfully as others, since the environment and other genes also have a say in the matter. How much offspring resemble their parents in a particular character is called the heritability of that character. Of course, the rate of natural selection is dependent not only on heritability, but also on the selective advantages and disadvantages of the behavior in question. Sometimes such unexpected factors as preferential mating among bearers of a trait brought about by the trait itself strongly influence the rate of natural selection. If the trait results in getting to the breeding grounds early, as it does in the blackcap, then the bearers of this trait will mate among themselves more often than they would by chance alone.

Let us therefore make the reasonable assumption that genes can influence a behavior enough that natural selection can then influence the rate of spread of that behavior. Armed with this assumption, let us see how natural selection influences different kinds of behaviors, especially in social animals.

4

....

What Do Social Animals Do to Each Other?

Logic suggests that when animals interact with each other, and social animals are more likely to do so, there can be four kinds of consequences. As a result of any social interaction, each partner may either benefit or suffer. For convenience, I shall refer to one of the partners, the one who initiates the interaction, as the *actor,* and the second, the relatively more passive one, as the *recipient.* Natural selection of course is blind to any benefit or cost unless it affects the reproductive fitness of the individuals concerned. I will therefore mean by *benefit* an increase in reproductive fitness and by *cost* a decrease in reproductive fitness.

You Scratch My Back and I'll Scratch Yours

When both the actor and the recipient benefit from an interaction, the interaction is *cooperative.* Cooperation is widespread in the living world, both within and between species. The apple I am eating while writing this is the result of cooperation between animals and plants. Many species of plants provide pollen and nectar for the benefit of bees, which in turn pollinate the plants. In the majority of cases the bees inadver-

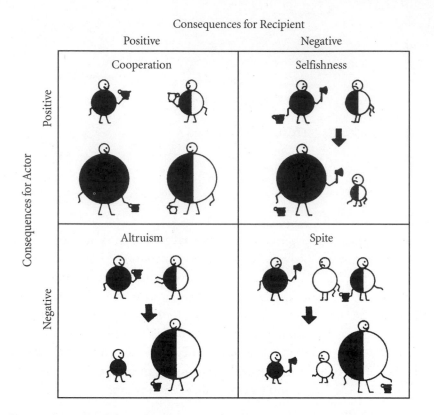

The consequences of interaction between animals. The recipient here is the actor's brother and therefore shares 50 percent of his genes, as is indicated by the shading. Help of any kind (the offering of food or shelter, easing access to a mate, and so on) is indicated by a vessel, and harmful behavior by an ax. *Cooperation:* Both individuals benefit and such behavior will therefore evolve easily. *Altruism:* The altruist diminishes his own genetic fitness but raises his brother's fitness to the extent that the shared genes are actually increased in the next generation. *Selfishness:* The selfish individual reduces his brother's fitness but increases his own to an extent that more than equal's the brother's loss. *Spite:* The spiteful individual lowers the fitness of an unrelated competitor (the unshaded figure) while reducing that of his own or at least not improving it; but the act increases the fitness of the brother to a degree that more than compensates for the actor's loss. *(Modified and redrawn with the permission of Harvard University Press from Wilson 1975.)*

tently drop pollen from their previous visits onto new flowers from which they are gathering fresh pollen or nectar. Notice that natural selection does not care whether the pollen was accidentally dropped or whether the bees deliberately placed pollen on the stigmas of the new flowers. As long as there is a high probability of being pollinated, plants that produce large quantities of pollen and nectar and have adaptations to attract insects have higher reproductive success than individuals that don't and will therefore be favored by natural selection. Similarly, the bees that go from flower to flower and efficiently collect large quantities of pollen and nectar have a higher reproductive success than bees that don't and will therefore be favored by natural selection.

Cooperation here is passive and the benefit accrued to the plant is almost a side effect. As long as it is pollinated, the plant does not care whether a particular bee benefits from its pollen and nectar. Similarly, as long as it gets pollen and nectar, the bee does not care whether a particular flower gets pollinated or not. Usually, plants and bees are not very species-specific and the plants will not suffer unless there is a general decline in bee populations. Similarly, bees will not really suffer unless there is a general decline in nectar- and pollen-yielding plant populations. I hope you see that "bees don't care whether a particular flower gets pollinated or not" is shorthand for the cumbersome statement that "bees that ensure that each flower providing them with pollen and nectar receives enough pollen to be fertilized do not usually have any higher fitness than bees that do not have any particular adaptations for ensuring such justice to flowers." When the welfare of each flower is in the interest of the pollinator, when insects that ensure justice to the flowers are fitter than those that do not, natural selection has produced appropriate adaptations.

The yucca plant and the yucca moth have entered into a very specific long-term evolutionary pact. Yucca flowers are pollinated by yucca moths. The moth does more than collect pollen and inadvertently pollinate flowers; it lays its eggs in the flowers so that its offspring feed on the seeds. But the moth larvae eat only a small proportion of the seeds and leave the rest intact so that the yucca plant can reproduce. In this case the moth has reasons to care about the individual flower that it has pollinated because it has invested the future of its own offspring in that particular flower. Sure enough, now pollination is no longer a passive process. The moth actively deposits enough pollen in the stigmas of the flowers so that its offspring have enough developing seeds to feed on. Why don't the moths lay more eggs so that all the seeds are eaten and more moth larvae develop? This would be foolish, because the plant would then go extinct. It turns out, however, that the yucca plant does not rely exclusively on such long-term checks on exploitive behavior by the moths. The plant routinely aborts flowers that have too many moth eggs and retains and nourishes only those that have a small number of eggs. The flowers that have no moth eggs at all of course are not pollinated and hence are useless to the plant.

Cooperation, whether within or between species, whether passive or active, is easy to understand because it benefits both parties. Like the yucca plant that aborts flowers containing too many moth eggs, however, nobody takes cooperation from the other party for granted; there is a constant effort at one-upmanship. Cooperation and conflict are inseparable.

Move Over, I'm the Boss Here

When the actor benefits and the recipient suffers, the interaction is *selfish*. Selfishness is even more abundant in the living world than cooperation. Indeed, selfish competitiveness is the cornerstone of the theory of natural selection. If you have ever watched a troop of monkeys you have seen that there is an easily detectable hierarchy among the members of the troop; this is especially evident among the males. The dominant male will readily accept any food you offer him and will have first priority in copulating with the females of the troop. The subordinate males are clearly afraid of the dominant male and will hesitate to accept any food you offer them while the dominant male is watching. Subordinate males will also only copulate stealthily when the dominant monkey is looking the other way. The dominant individual is selfish in usurping the choicest food and the best mating opportunities for himself. This behavior enhances his reproductive fitness and may be quite detrimental to the subordinates. But natural selection will have no difficulty in favoring such selfish behavior on his part, the rule being survival of the fittest, after all.

I'm Hungry But I'm Sure You Are Even Hungrier

In the 1970s Paul Sherman, then at the University of California, conducted a unique study of the alarm calls of Belding's ground squirrels. These are diurnal rodents that inhabit the alpine and subalpine meadows of the western United States. During an 8-year period a number of squirrel enthusiasts participated in a project to individually mark 1866 squirrels in Tioga Pass Meadow, in the Sierra Nevada mountains of California. The squirrels were marked by toe clipping or ear tagging for firm identification upon capture, as well as with human hair dyes for identification at a distance. Sherman watched these marked squirrels with a specific interest in their response to predators. The squirrels in this area have at least five predators: weasels, badgers, dogs, coyotes, and pine martens. Sherman spent 3082 hours observing and witnessed 102 occasions when a squirrel encountered or noticed a predator—less than once for every 25 hours of observation.

Like many other squirrels, Belding's ground squirrels give alarm calls at the approach of a predator. The alarm calls in response to terrestrial

A Belding's ground squirrel giving an alarm call. *(Drawing: Prema Iyer.)*

predators are in the 4–7 kilohertz range and are repeated 6–7 times per second. Giving alarm calls is certainly a risky proposition; it increases the probability of the callers being attacked by the predator by about two and a half times. Then why do the squirrels give alarm calls? Why don't they simply run away and hide and benefit from the fact that they have seen the predator before the other squirrels have? The individual that gives the alarm calls is behaving in a manner that decreases its fitness (its probability of survival and hence its reproductive success) and increases the fitness of others.

When the actor, suffers and the recipient benefits, the actor's behavior is *altruistic*. Altruism is less common than cooperation and selfishness. Nevertheless, there are many cases of altruistic behavior in the animal kingdom, such as the alarm-calling behavior of the Belding's ground squirrel; of course the sterile worker bee is a prime example of altruism. Such altruism is not easy to explain. One might say that altruism is paradoxical because altruists should leave behind fewer offspring and should lose out in the struggle for existence. Without a significant modification of the theory of natural selection, altruism cannot be explained; later I will discuss such a modification.

I'll Kill You Even If I Have to Starve

If both the actor and the recipient suffer as a result of an interaction, it is called *spite*. Conventional wisdom has it that to be spiteful is the prerogative of humans and that animals are not known to be spiteful. But by slightly relaxing the definition of spite to include those cases where the actor neither benefits nor suffers, but the recipient clearly suffers, we may be able to uncover examples of possible spite in the animal kingdom. The most clear-cut demonstration of such weak spite is seen in a fish, the three-spine stickleback. Sticklebacks have been a favorite species for behavioral studies on fish, perhaps because their most striking feature is their extensive cannibalism, especially their tendency to devour the eggs of other members of their own species.

The late Gerald J. FitzGerald performed a series of experiments to test if egg-eating sticklebacks can be called spiteful. In his area of study, near Isle Verte in Quebec, three-spine sticklebacks coexist with a related species, the black-spotted stickleback, which is more common than the three-spine form. Three-spine stickleback females have a preference for attacking and eating eggs of other members of their species even though eggs of the other species are available and are more common. Moreover,

A three-spine stickleback female eating another stickleback's eggs. *(Drawing: Sanjeeva Nayaka.)*

they seem to have a preference for eating eggs of members of their own population rather than those of different populations. Now there is evidence that they can distinguish eggs of the other species from eggs of their own species. And of course, by eating eggs of their own species, they run a significant risk of accidentally eating their own eggs because females deposit eggs in nests made by males and do not remain there to care for their young. What can explain this behavior? Is it possible that they prefer to eat eggs of other members of their species instead of eggs of the related, more abundant, black-spotted sticklebacks because the former are nutritionally more valuable? In carefully controlled experiments FitzGerald showed that this is not true. There is a further curious aspect of their behavior which suggests that they might indeed be spiteful. Young eggs are clearly more valuable nutritionally than older eggs, but the older eggs are more valuable to their parent's fitness because they will hatch sooner. The three-spine stickleback females have a distinct preference for eating older eggs (less valuable to them but more valuable to the egg's parents) rather than younger eggs (more valuable to them but less valuable to the egg's parents) of other members of their species.

Are the three-spine sticklebacks behaving in a spiteful, or at least weakly spiteful, manner? Perhaps. Then how has natural selection favored such a behavior? We don't quite know the answer to that. No simple modification of the theory of natural selection that might accommodate spite has yet been formulated. If many more examples of spite are discovered, I believe that we will begin to carry such a large burden of unexplained facts that the motivation to modify the theory will become very strong. The problem is that any potential example of spite is rapidly dismissed by hypercritical referees of scientific papers with the result that we are in danger of overlooking such examples and perhaps delaying the development of an interesting new direction of inquiry in evolutionary biology.

5

The Paradox of Altruism

Darwin's Insuperable Difficulty

As we have already seen, the sterile worker bee strives to rear the queen's brood and usually dies without herself reproducing. And the worker bee is exquisitely adapted to perform her tasks. She has wax glands in her abdomen, pollen baskets on her hind legs, and the ability to perform an elaborate dance language to recruit nestmate workers to new sources of food. The queen bee has none of these abilities. How can natural selection favor the sterile honey bee worker that leaves behind no offspring or even the squirrel that reduces its chances of survival by giving an alarm call upon seeing a predator? More paradoxical, perhaps, how can the process of natural selection help perfect the adaptations of the sterile worker bee? We couldn't say, for example, that workers who had better pollen baskets left behind more offspring and gradually replaced those workers who had inferior pollen baskets. Only the queen reproduces and natural selection can only act on her.

It is a tribute to the genius of Charles Darwin that these questions bothered him, but I do not believe that he had a satisfactory answer. In *On the Origin of Species* Darwin referred to the worker honey bee as a "special difficulty, which first appeared to me insuperable, and actually fatal to my whole theory. I allude to the neuters or sterile females in insect communities: for these neuters often differ widely in instinct and in structure from both the males and fertile females, and yet, from being sterile, they cannot propagate their kind." Two paragraphs later Darwin summarizes his solution to the problem: "This difficulty though appearing insuperable, is lessened, or, as I believe disappears, when it is remembered that selection may be applied to the family, as well as to the individual and may thus gain the desired end. Thus a well-flavored vegetable is cooked, and the individual is destroyed; but the horticulturist sows seed of the same stock, and confidently expects to get nearly the same variety; breeders of cattle wish the flesh and fat to be well marbled together; the animal has been slaughtered, but the breeder goes with confidence to the same family."

Historians have now suggested that Darwin's analogy with artificial selection of cattle does not ring true; perhaps it didn't quite satisfy Darwin himself. Darwin delayed the publication of his theory of natural selection for years and might have delayed it even further if Alfred Russel Wallace had not independently hit upon the idea and thereby spurred Darwin to publish. Writing in the *Journal of the History of Biology*, F. R. Prete has described "the conundrum of the honey bees" as "one impediment to the publication of Darwin's theory." Prete's point is that unlike the queen and worker bees, the slaughtered cow and the cattle used for breeding are both almost identical and that is why the breeder goes with confidence to the same family. The worker bee is quite different from both the queen and the drone, and yet nature appears to go with confidence to the queens and drones of hives containing workers with superior pollen baskets to get more workers with superior pollen baskets. To make Darwin's analogy with cattle apply to the bees, we have to imagine, in Prete's words, that "a cattle breeder has a skinny pure white cow and an ugly pure black bull. When bred, these animals invariably give rise to large herds of beautiful, brown, quality beef cattle, all of one sex, and an occasional pair of breeders (one skinny white and the other ugly black) who could repeat the process." Not surprisingly, Prete concludes that "it is highly improbable that Darwin, as insightful and meticulous as he was, did not also consider this difficulty."

When Would You Risk Your Life to Save a Child?

John Burdon Sanderson Haldane was a truly remarkable man. John Maynard Smith, his student and colleague, writes of him: although "Haldane will be remembered for his contribution to the theory of evolution . . . he is in other respects somewhat difficult to classify. A liberal individualist, he was best known as a leading communist and contributor of a weekly article to the *Daily Worker*. A double first class in classics and mathematics at Oxford, he made his name in biochemistry and genetics. A captain in the Black Watch who admitted to rather enjoying the First World War, he spent the end part of his life in India writing in defense of non-violence."

In an obscure little article that appeared in 1953 in a now defunct magazine called *Penguin New Biology*, Haldane sowed the seeds of an idea that provides a satisfactory solution to Darwin's insuperable difficulty. Before I read Haldane's article in the original, I had heard of it in the form of a story which goes something like this: Haldane was once walking on the bank of a river with a friend. As was typical of him, Haldane paused for a moment, made a quick calculation on the palm of his hand, and declared: "If one or two of my brothers were drowning in this river, I might perhaps not risk my life to save them but if more than two of my brothers were drowning, I might attempt to save them at a

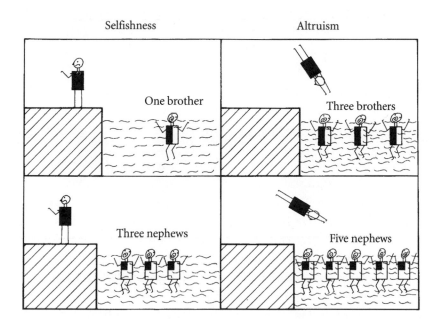

Cartoon illustrating the theme of J. B. S. Haldane's story. The shaded portions of the drowning individuals indicate the proportion of their genes which are also present in the altruist standing on the bank. Notice that the altruist is willing to risk his life when the numbers of his genes expected to be rescued is greater than the number in his body likely to be lost. *(Drawing: Sudha Premnath.)*

risk to my life." The story may be pure fiction, but I find it very useful in teaching students the modern solution to Darwin's paradox. And it no doubt faithfully reflects Haldane's written version, at least as far as the scientific idea is concerned. Haldane wrote, "Let us suppose that you carry a rare gene which affects your behavior so that you jump into a flooded river and save a child, but you have one chance in ten of being drowned . . . If the child is your own child, or your brother or sister, there is an even chance that the child will also have this gene, so five such genes will be saved in children for one lost in the adult. If you save a grandchild or nephew the advantage is only two and a half to one. If you only save a first cousin, the effect is very slight. If you try to save your first cousin once removed the population is more likely to lose this valuable gene than to gain it."

The Concept of Inclusive Fitness

W. D. Hamilton has given us the required modification of Darwin's theory that can accommodate altruism. Hamilton argued that altruism is no paradox at all if we realize that natural selection is dependent on changes in the relative frequencies of genes (alleles) regardless of the pathway by which the change is brought about. What this means of course is that producing offspring is only one way to increase the representation of one's genes in the population. Aiding genetic relatives who carry copies of one's genes is another, equally legitimate, way of doing so. To put it more starkly, you can be sterile and still have fitness. But how do we decide whether a sterile individual is just as fit as or fitter than a fertile one? We have no difficulty in deciding that an individual producing two offspring is fitter than another producing only one offspring. But how do we compare the fitnesses of individuals producing one offspring and those devoting their lives to taking care of, say, one brother or three cousins or five nephews.

This is where we can go back to Haldane's logic. On the average, we share one half of our genes with our offspring and siblings, one fourth with our grandchildren and nieces and nephews, one eighth with our cousins, and so on. As far as evolution is concerned, caring for one child is equivalent to caring for one sibling, or two grandchildren, or two nephews, or four cousins, and so on. Genetically speaking, we can express any class of relatives as offspring equivalents and then compare the fitnesses of individuals with different propensities for rearing offspring or aiding relatives. Hamilton went a step further and converted everything into genome equivalents (a genome being the entire genetic material of one individual). This is easily done by multiplying the number of offspring and siblings by 0.5, the number of grandchildren and nieces and nephews by 0.25, the number of cousins by 0.125, and so on. The contribution of different classes of relatives to fitness can then be added up to yield the *inclusive fitness*. Now we can appreciate Haldane's reluctance to risk his life to save only one or two brothers and his readiness to risk his life to save three or more brothers. Being related to himself by 1.0, he would need to save three or more brothers (0.5 $\times$ 3 or more) to make up for the loss of his entire genome.

Consider another example. Praveen Karanth and S. Sridhar studied the breeding behavior of the small green bee-eater in and around Bangalore, India. They found that in about 40 percent of the birds' nests there was a helper in addition to the breeding pair. The helpers must truly help because nests with helpers produced more fledglings per nest than nests without helpers, and these fledglings grew more rapidly and had fewer problems with predation than fledglings without helpers. Karanth and Sridhar did not know the genetic relationships between the helpers and the breeding pairs. But we know from other species of birds with the helping habit that older offspring often help their parents rear a second brood. Suppose a young bee-eater that goes off to breed can produce two chicks while one that stays to help its parents contributes to the survival of three more chicks than the parents can raise without help. The inclusive fitness of the helper will be greater than that of the one that goes off to breed on its own. Notice, however, that we should credit helpers only with the additional chicks reared because of their help, and not assign fitness to them for the chicks that might have survived anyway. When we were only counting offspring it was easy to assign credit. When we also count relatives in assigning fitness, there is a danger of double counting and we must guard against it.

The Two Components of Inclusive Fitness

I study a socially primitive wasp called *Ropalidia marginata*. These wasps form social colonies with nonreproducing workers and reproductive queens. Queens and workers are not morphologically different and hence individual wasps can act as queens or workers in response to the opportunities available. I have come across some individuals who work for a time, helping their mothers to produce more offspring, and later drive out their own mothers and become queens in the same colony. How do we compute the inclusive fitness of such individuals pursuing multiple strategies? That's simple enough. We can just convert everything into genome equivalents and then add up the fitness gained through offspring and that gained through relatives. Inclusive fitness, then, has two components, a direct, individual component, gained through selfish, offspring production, and an indirect, social component, gained through altruistic caring for genetic relatives. The sum of these two components is what matters, and therefore even if one component is zero the sum may still be very large. That then is the secret of the evolutionary success of sterile honey bee workers.

It is worth emphasizing that natural selection does not, in any way, break up inclusive fitness into direct and social fitness components. Indeed, natural selection cannot distinguish between fitness gained through the direct component and fitness gained through the social component, and that is why two individuals with the same level of inclusive fitness are identical in the eyes of natural selection even though one may have gained all of its inclusive fitness through the direct component while the other may have gained it all through the social component. Then why should we break inclusive fitness up into direct and social components? Because if we want to measure the inclusive fitness of animals in real life, it is convenient and even necessary to do so. The circumstances and the strategies associated with accumulating direct fitness are often different from those associated with the acquisition of social fitness. Also, in the case of the social component, one has to worry much more about the level of genetic relatedness between the actor and the recipient, while in the case of the direct component, the relationship between parent and offspring is almost always 0.5. When a behavior is favored by natural selection exclusively or primarily because of its contribution to the social component of inclusive fitness, the behavior is said to have evolved by *kin selection.*

Hamilton's Rule

We have just derived a fundamental rule in evolutionary biology that is known as Hamilton's Rule. Stated in more technical terms, Hamilton's rule is that an altruistic trait can evolve if the number of individuals gained, multiplied by the altruist's genetic relatedness to those individuals, is greater than the number of individuals lost, multiplied by the altruist's relatedness to them. If Haldane had rescued three brothers and lost his life in the process, the number of individuals gained multiplied by his relatedness to them ($3 \times 0.5 = 1.5$) would have been greater than the number of individuals lost, namely Haldane himself, multiplied by Haldane's relatedness to himself, 1.0 ($1 \times 1.0 = 1.0$). Thus the altruistic trait of risking one's life to save some one in danger *can* evolve by natural selection, provided of course that other conditions such as a genetic basis for the behavior are met. If an altruistic bee-eater helps its parents produce an additional three sibling chicks, its inclusive fitness as a helper is the number of individuals gained times its relatedness to

Hamilton's Rule

$$b/c > 1/r$$

b = benefit to recipient
c = cost to donor
r = genetic relatedness between donor and recipient

or

$$n_i\, r_i > n_o\, r_o$$

n_i = no. of relatives reared
r_i = relatedness to relatives
n_o = no. of offspring reared
r_o = relatedness to offspring

Hamilton's rule defines the condition for the evolution of altruism. The upper form is useful to predict when an individual will be selected to sacrifice its life to help others. The lower form is useful to predict when a sterile individual who rears relatives will be selected over a fertile individual who rears offspring.

them: $3 \times 0.5 = 1.5$. If the helper had produced two offspring instead, its inclusive fitness would be the number of individuals gained multiplied by its relatedness: $2 \times 0.5 = 1.0$. Its inclusive fitness as a helper would be greater than its inclusive fitness as a breeder and hence Hamilton's rule is satisfied and the altruistic trait can evolve. Notice that we can only say that it *can* evolve; we cannot assert that it *will* evolve. The trait can only evolve if other conditions such as its having a genetic basis are met. Our strategy will thus be to see what types of behaviors *can* evolve and what types *have* actually evolved. If what can evolve has evolved, that will be satisfying because it will suggest that we are on the right track in our theorizing. If we find that what can evolve has not evolved, and especially if what cannot evolve according to theory has evolved, we will be challenged to refine our theory.

At the risk of stating the obvious, let me stress that Hamilton's rule does not just provide a theory for the evolution of altruism. It simultaneously and automatically provides a theory for the evolution of selfishness. If Haldane had risked his life to save just one brother, his inclusive fitness ($1 \times 0.5 = 0.5$) would have been less than it might have been if he had looked the other way ($1 \times 1.0 = 1.0$). If a bee-eater foregoes an opportunity to breed and becomes a helper but can only increase the number of fledged chicks in its parents' nest by one, it will similarly suffer a net loss in inclusive fitness. Under these conditions, selfishness is expected to evolve rather than altruism. Cooperation, of course, is easily explained by Hamilton's rule because it will increase the inclusive fitness of the actor without any associated cost (both actor and recipient benefit).

There is, however, a problem that at first sight appears incapable of explanation by Hamilton's rule. If we are equally related to our offspring and our siblings, why is care and attention showered so much more generously on offspring than on siblings throughout the animal kingdom? Hamilton's rule by itself appears to predict no particular preference for one over the other. But that is not true, because here we are only looking at the identical relatedness to offspring and siblings. We must also focus on the benefit and cost, and here there may be quite a difference between offspring and siblings. Animals routinely have access to young and helpless offspring at a time when they themselves are adults. The cost of helping offspring in this situation will be relatively small but the benefit to the offspring will be great. The situation is different with siblings: they are often roughly the same age and at the same level of helplessness and therefore the cost of their helping each other will be high and the benefit not so great. Thus Hamilton's rule does predict that help will be given more often to offspring than to siblings.

Reciprocal Altruism

There is yet another way altruism can evolve. Imagine you are hungry today but have no money to buy food. And if you don't eat you might actually die. I have a little more money than I need for today, but I better not give it to you because I may end up like you tomorrow, and you are not even related to me. But of course I might go hungry and be broke tomorrow even if I don't give you the extra money I have today. On second thought, I will give you the extra money I have today. Perhaps some day I will be hungry and broke and you will remember my good deed and help me out. On that day you will probably not suffer greatly by giving me your extra money, but it will make all the difference between life and death for me just as it will for you today. We will both benefit from such *reciprocal altruism.* We might both have died if we had not helped each other. Reciprocal altruism has made it possible for both of us to survive. You will see no doubt that my helping you would not have been a good idea if there was not a high probability that you would return the favor when I needed it. Thus reciprocal altruism can evolve if (1) there is a good chance that the same actors will meet repeatedly; (2) they can recognize each other; and (3) they can remember who helped them in the past and who did not. The last-mentioned condition is of utmost importance because reciprocal altruists need to guard against cheaters. Of course it is best for me to take your help when I am in trouble and avoid helping you when you are in need. Cheaters can go scot free if the conditions of repeated encounters, recognition, and memory are not met. But then the trait of helping will not pay dividends and hence will not be favored by natural selection.

The idea of the evolution of altruism by reciprocity was proposed by Robert Trivers. But do animals practice reciprocal altruism? There is at least one good example, and strangely enough it also has to do with donating extra food in times of excess to those in need. The only difference is that it concerns the sharing of extra blood by vampire bats,

The common vampire bat *Desmodus rotundus* feeding on blood provided in plastic trays in the laboratory. *(Photo: Merlin D. Tuttle, Bat Conservation International.)*

not quite what I might have given you when you were hungry. Gerald Wilkinson studied vampire bats in Costa Rica. These bats live in groups of 8 to 10 females, some of whom are sisters but some of whom are unrelated to each other. These groups of females associate with each other for 2 to 11 years and thus they have ample opportunities to indulge in reciprocal altruism. Vampire bats fly out at night to feed on the blood of cattle and horses and then return to their roosting sites to spend the day. Not all bats succeed in feeding on all nights but who succeeds and who fails appears to be a matter of chance. Bats that fail to feed on three consecutive nights will almost certainly die of starvation. Wilkinson found that hungry bats will beg food from well-fed ones and will usually be offered some blood. Bats receiving blood are more likely to donate blood when they themselves are well fed and are importuned by hungry bats. The bats groom each other on their stomachs and this appears to be their way of telling who is well fed and who is hungry. There is evidence from laboratory studies that they can remember the individuals to whom they have donated blood in the past. These bats can ingest an amount of blood equal to their body weight and hold most of it in highly distensible stomachs; this must make it very difficult for a well-fed bat to pretend, especially while being groomed on its stomach, that it has nothing to regurgitate. Thus there appears to be a reasonable mechanism to detect and punish cheaters—an essential condition for the evolution of reciprocal altruism.

Is It All Just Selfishness?

One solution I offered for the paradox of altruism is that it is no paradox at all if altruism is directed toward genetic relatives and the net gain due to the increased survival of copies of the altruist's genes through the helped relatives is greater than the loss of copies of the altruist's genes due to its own death or a reduction in the number of its offspring. In short, apparent altruism at the level of the individual animal is no altruism at all at the level of the genes; it's really selfishness from the point of view of the genes. The second solution I offered for the apparent paradox of altruism is that altruism may be practiced because of the high probability of its being reciprocated when the giver may be in more need of help than it is now. You might argue that this is no altruism either but is instead a very clever kind of selfishness. So is all altruism really selfishness? Perhaps it is.

Many people find this conclusion very unsatisfactory. Some argue that calling altruism selfishness at some other level or in a longer time frame is no way of solving the paradox of altruism. I don't agree; that may just be what it is and there may be no altruism at the level of the genes. Selfishness at the level of the genes can lead to several behavior patterns, including selfishness, cooperation, altruism, or even spite at the level of the individual organism—whichever leads to better selfishness at the level of the genes. Other people argue that at the very least we must stop calling the behavior altruism when we realize that it is a hidden form of selfishness. I don't agree. To us in our day-to-day experience, the individual animal is an obvious entity, and if individuals show altruism it is worth distinguishing it from selfishness at the individual level even if it is selfish at the gene level. Similarly, if animals show altruistic behavior that is reciprocated after a significant time delay, it is worth distinguishing it from routine selfishness. Nothing is gained by labeling everything selfishness. It is only because behaviors recognized as altruistic and apparently paradoxical received so much attention that they engendered in Darwin himself uncertainty about his theory and, later, led to its modification in the form of inclusive fitness theory. Besides, labeling all altruism toward relatives and all reciprocal altruism as selfishness would amount to reserving the term "altruism" for anything that cannot evolve—because we really have no theory to explain altruism if it is not directed toward genetic relatives and if it is not reciprocal, unless of course we uncover genuine examples of the evolution of altruism by group selection.

6

Do Animals Favor Their Relatives?

I trust it is obvious by now that terms such as "altruism," "selfishness," and "spite" are routinely used in studies of animal behavior and evolutionary biology for the sake of convenience. They mean approximately what they mean in the human context that we are all familiar with, but they are objectively and precisely defined in terms of the fitness consequences to the actors and recipients. Recall that in Chapter 4 we defined cooperation as an interaction where both the actor and the recipient benefit, selfishness as an interaction where the actor benefits while the recipient suffers, altruism as an interaction where the actor suffers while the recipient benefits, and spite as an interaction where both the actor and the recipient suffer. We have also seen that natural selection is blind to any cost and benefit unless it affects the reproductive fitness of the actors and recipients concerned.

When animals favor close genetic relatives over nonrelatives or distant relatives as recipients of beneficial acts, they are said to be practicing *nepotism.* The use of the term "nepotism" in animal studies does not involve any moral connotation, as it almost always does when applied to humans. When we describe acts of altruism, bravery, or chivalry

81

among humans, there is usually at least an implicit nod of appreciation and approval, and when we describe acts of selfishness, nepotism, or spite among humans, there is usually a connotation of disapproval. This judgmental attitude is what we must scrupulously avoid when dealing with animal behavior. We must discipline ourselves to ask whether animals are altruistic without conveying any sense of approval, and we must ask if animals are nepotistic or spiteful without attaching any negative connotations to those terms. We must even be prepared to ask if it is "good" for an animal to be nepotistic, because by "good" we mean whether the nepotism increases its biological fitness. This use of language has nothing to do with whether it is good or bad for humans to be selfish, altruistic, nepotistic, or spiteful. We should decide for ourselves what is good and what is bad for humans; human ethics should not be based on what animals do or don't do. As Richard Dawkins wrote in the forward to Robert Axelrod's book on the evolution of cooperation "If you wish . . . to build a society in which individuals cooperate generously and unselfishly towards a common good, you can expect little help from biological nature. Let us try to *teach* generosity and altruism, because we are born selfish. Let us understand what our selfish genes are up to, because we may then at least have a chance to upset their designs, something that no other species has aspired to do."

Let us therefore turn to the animals to satisfy our curiosity about what animals do and our urge to understand the mysteries of nature but not to find justifications for our ideas about how humans should behave. Some people have argued that we should not use words such as "altruism" and "nepotism" but should invent new words for the same behaviors in animals. I don't believe that would help either the cause of animal behavior studies or the cause of human ethics. It would make communication of science to nonspecialists and, indeed, even communication among specialists, even more difficult, and it would not inhibit those who want to use the findings of animal behavior to justify their philosophy of human behavior. Only a conscious decision not to blindly draw moral lessons from what animals do can help in this matter.

Some years ago there was an interesting exchange of letters in the science magazine *Nature*. It all started with a photograph of the social wasp *Polistes* that appeared on the cover of one of the issues of *Nature* accompanied by a caption describing the altruism of the wasp workers. Raymond Rasmont, a biologist from Belgium, wrote a letter to the editor of *Nature* objecting to the use of the word "altruism" to describe the behavior of the wasps. His objection was that a moral judgment is inevitable when the same word is used in a human context and we do not wish to imply any such moral judgment to the behavior of the wasps. Rasmont therefore suggested that biologists should accept that altruism has long been used with an ethical connotation and should coin a new term for animals, "as free from ethical connotation as possible." He suggested the word "euxeny," derived from a Greek word meaning to be nice to strangers. I was not entirely surprised at this letter since such debates have raged for years. But I was a bit surprised that the objection had come from a biologist; I would have expected a sociologist to claim priority for the use of the word "altruism" in the human, ethical context. I was therefore even more surprised to see a couple of months later another letter to the editor of *Nature*, this time criticizing Rasmont, that had been written by a social scientist, Christopher Badcock, from the London School of Economics. Badcock pointed out that what appears to be altruism on the part of the animals is, as we have seen, really selfishness at a deeper level. To keep "euxeny" distinct from human altruism would, Badcock said, "really make it too easy for the politicians" and lend credence to politicians' claims that they always "act in the public interest, rather than in their own self-interest." And he went on to note that Anna Freud's classic study of the motives involved in human altruism concludes that "it remains an open question whether there is such a thing as genuinely altruistic relation to one's fellow men, in which the gratification of one's own instinct plays no part at all."

Why Do Bee-Eaters Help?

We saw earlier that small green bee-eaters in Bangalore sometimes help other breeders rather than breed on their own. This behavior is known to occur in a number of different kinds of birds, such as the Florida scrub jay, the Galápagos mocking bird, the jungle babbler, the acorn

The small green bee-eater *(Merops orientalis)*. *(Photo: S. Sridhar.)*

woodpecker, the pied kingfisher, the splendid wren, and the white-fronted bee-eater, to name just a few. Stephen Emlen and Peter Wrege have attempted to find out why bee-eaters help rather than breed on their own. They studied the white-fronted bee-eater, which is a close relative of the small green bee-eater, for several years in the Lake Nakuru National Park in Kenya. The white-fronted bee eater lives in extended family units, or clans. In each breeding season, numerous clans aggregate to form large colonies of about 200 birds each. About 50 percent of all nests have at least one nonbreeding helper that participates in excavating and defending the nest, feeding the breeding female before she lays eggs, incubating the eggs, and feeding the nestlings and fledglings.

The white-fronted bee-eater *(Merops bullockoides). (Photo: Natalie J. Demong.)*

Every year, Emlen and Wrege painstakingly sexed and individually marked the birds for identification, and also recorded for every nest the number of helpers, the identity of the breeders and that of the helpers, the clutch size, the hatching success, and the fledging success. Through their lengthy observation of the birds and their behavior, Emlen and Wrege were able to determine the genetic relationship between the helpers and the helped. With the resulting data, they were able to test nine different hypotheses that had been proposed to explain how helpers might benefit from helping.

These hypotheses may be broadly divided into four classes: (1) helpers have a better chance of survival if they associate themselves with a nest; (2) by helping, helpers increase their probability of becoming breeders in the future; (3) helpers are more successful at breeding if they have had past experience as helpers; and (4) helpers gain fitness indirectly through helping relatives. Notice that the first three classes of hypotheses depend on direct benefit to the helpers—the individual component of inclusive fitness. The fourth class of hypotheses depends on indirect benefit via relatives—the social component of inclusive fitness.

Emlen and Wrege found no evidence that an individual increases the chances of its survival by being a helper, nor did they find that helpers are more likely to become breeders in the future. If helpers do become breeders, they are not any more successful than birds without prior experience as helpers. But Emlen and Wrege did find clear evidence that helpers help only when close relatives are available to receive the help and that when they do help they significantly increase the survival of the nestlings. In other words, there is clear evidence against the first three classes of hypotheses involving the individual fitness component and equally clear evidence in support of the fourth class of hypotheses involving the social component of inclusive fitness. The white-fronted bee-eaters are altruistic, and that is not paradoxical from an evolution-ary point of view because they are nepotistic in the dispensation of their altruism; they help because it enhances their inclusive fitness via the social component. Natural selection will of course favor helping behavior as long as it increases their inclusive fitness, whether through the individual component or through the social component. In this case, it just happens that the increase in inclusive fitness comes through the social component.

Why Do Squirrels Give Alarm Calls?

We saw earlier that Belding's ground squirrels in California give alarm calls at the approach of a predator to warn other members of their population, at a significant risk to themselves. Now why do they do that? How would natural selection favor such altruistic behavior? The theory developed in Chapter 5 suggests that altruism would be favored by natural selection either if it is directed toward genetic relatives or if it is later reciprocated by the beneficiaries. You will recall that Paul Sherman had 1866 squirrels marked for individual identification and spent 3082 hours observing them, during which he witnessed 102 occasions when the squirrels encountered predators. The aim of Sherman's study was to discriminate between six competing explanations of the possible function of squirrel alarm calls: (1) distracting the predator's attention, (2) reducing the likelihood of future attacks by the same predator, (3) discouraging the predator by indicating that it has been spotted, (4) warning others likely to reciprocate, (5) warning the group as a whole, and (6) warning genetic relatives.

Sherman found no evidence that warning calls distracted the attention of predators or reduced the likelihood of future attacks. Nor did he find any evidence that warning calls discouraged predators from pursuing the caller; indeed, callers were often chased and pursued and even killed by predators. That rules out the first three hypotheses. Hypotheses 4 and 5 were more difficult to falsify. There appeared to be no difference in the probability of giving alarm calls when temporary invaders (who are not likely to be reciprocators) were present and were not present. It is therefore unlikely that alarm calls have evolved by the system of reciprocal altruism discussed in Chapter 5. This makes hypothesis 4 unlikely. If alarm calls had evolved by a process of group selection, then, at the very least, one should find that some groups of squirrels give alarm calls while others do not. But Sherman found no evidence that alarm callers were restricted to some groups. He therefore did not favor hypothesis 5. In contrast to all these negative results, Sherman found that squirrels are most likely to give alarm calls when their living relatives are present in the area. Thus the only hypothesis that was supported was hypothesis 6, that alarm calls serve to warn relatives of impending danger. As we saw earlier, alarm calls can be favored by natural selection even if they reduce the fitness of the caller if they also benefit relatives. Sherman's conclusion is reflected in the title of his research paper: "Nepotism and the Evolution of Alarm Calls."

Why Do Slime Molds Commit Suicide?

We saw earlier that some soil amoebae commit suicide by differentiating into dead stalk cells so that others can differentiate into spores and disperse efficiently to new and better habitats. Why should natural selection favor such behavior? An amoeba with a selfish genetic constitution, or genotype, would obviously prefer to have its cells become spores rather than stalk cells. Such selfish genotypes should be favored by natural selection until ultimately all cells are selfish. If this means that no cell disperses to better habitats and that they all die, so be it—natural selection will not change things, unless altruistic stalk cells can somehow gain fitness.

Almost nothing is known about how slime molds live in nature; most of what we know about them comes from experiments in the laboratory. Experimenters prefer to start their laboratory culture of slime molds with a single spore so that all the cells are descendants of a single parent cell. In other words cells that commit suicide to aid others in the petri dishes in the laboratory are all clones of each other. Now it does not matter which cells become stalk cells and which become spores because genetically they are all the same. Natural selection should favor a genotype that uses some of its cells to build a stalk and aid others to disperse. There is no question of selfishness on the part of the stalk cells because they are genetically identical to the cells that make the spores. This is something like our using some cells in our body to perform the functions of respiration and digestion so that the sperm and eggs can concentrate on making new individuals. We don't normally say that our intestinal cells are committing suicide to aid the sperm cells to disperse, do we? But if the intestinal cells are even slightly genetically different from the sperm or egg cells, natural selection should, according to theory, favor selfish cells that try to become sperm rather than intestinal cells.

Of course, we cannot easily test this hypothesis with the cells of our body. This is where slime molds come in handy. If we persuade amoebae that are not genetically identical to come together to form a slug, then we should see selfishness and competition to form spores rather than stalk cells. This is just the experiment that M. J. DeAngelo and several colleagues performed some years ago. They mixed two genetically different strains of slime mold and found that, compared with single-strain cultures, the mixed-strain cultures clearly exhibited competition to form spore cells. The ratio of spore to stalk cells went up significantly, so that small stalks and large spore masses were produced. This may have been bad for dispersal, but as I have said before, natural selection usually has no way of controlling such selfish behavior, at least in the short run. The selfishness evident in the mixed-strain cultures inspires confidence in the idea that altruism on the part of stalk cells in laboratory cultures has to do with the clonal nature of these cultures. So, why do slime molds commit suicide? The answer seems to be that they do so to help their relatives.

Why Do Tasmanian Hens Have Two Husbands?

The Tasmanian native hen, a flightless bird, is remarkable in being polyandrous—breeding groups consist of one female and one or more males. The female probably benefits from having more than one male around but why should the males tolerate each other? The question can be asked in two different ways. Why should a male, already in possession of a female, permit another male to join them, and why should a male looking for a female join one already paired with another male? The second question is easy to answer. For reasons that we need not go into, there is a shortage of females so that wives have to be shared. A new male looking for an opportunity to breed may have no choice but to accept a female that is already paired. But why the original male accepts the second male is a more difficult question. Needless to say, the males fight. One of them is usually more dominant than the other. So it is more appropriate to ask why the dominant male tolerates the subordinate one rather than to talk about the first and second males; the order in which they approach the female is probably less important than their respective strengths. Let us attempt to apply Hamilton's rule here. The benefit to the subordinate male of being allowed to stay must be about half the chicks that the trio will produce because both are known to copulate with the female. There is no cost to his staying because, if he does not, he seems to have no chance at all of finding an unattached female. So the subordinate male will always want to stay. But what about the dominant male? By allowing the subordinate to stay the dominant suffers a cost. This is equal to the difference between the number of chicks that he might have produced without the help of the subordinate and half the number that the trio will produce (recall that only about half the number produced by the trio will be his; the other half will be sired by the subordinate).

Let us take a real-life example. John Maynard Smith and M. G. Ridpath found that if you consider experienced breeders, a pair will produce, on average, 5.5 chicks per season while a trio will produce 6.5 chicks per season. That the trio does better than the pair is not surprising because the trio will be better able to provide for and protect the chicks. The benefit to the subordinate male of being allowed to stay is thus half of 6.50, 3.25 chicks. The dominant male would have produced 5.50 chicks if he had driven away the subordinate and has to be content with his share of 3.25 chicks if he allows the subordinate male to stay. His cost of accepting the subordinate is thus $5.50 - 3.75 = 1.75$ chicks. Tolerating the subordinate is thus an act of altruism at least from the dominant individual's point of view, if not from the point of view of his genes. Now, according to Hamilton's rule, such altruism on the part of the dominant male can evolve if the benefit to the recipient (the subordinate male) multiplied by the genetic relatedness between the dominant male and the subordinate's chicks is greater than the cost to the dominant multiplied by the genetic relatedness between the dominant male and his own chicks. The problem, however, is that Ridpath did not know whether the males sharing a female were related.

Maynard Smith and Ridpath first considered the possibility that the males are unrelated to each other. In that case the genetic relatedness between the dominant male and the subordinate's chicks will be zero. Thus altruism cannot evolve unless the cost to the dominant male is also zero. But there is always a cost associated with tolerating an extra male. The conclusion then is that altruism will not evolve in this system unless the cooperating males are related. Maynard Smith and Ridpath then considered the possibility that the males sharing a female are brothers. If so, the offspring of the subordinate male will be nieces and nephews of the dominant male and thus related to him by 0.25, while his own offspring will of course be related to him by the usual 0.50. Thus, altruism can evolve if one fourth of the benefit is greater than half the cost: the benefit of the altruistic act is the number of nieces and nephews raised and they are each related to the dominant male by 0.25, while the cost of the altruistic act is the number of offspring given up and they would each have been related to the dominant male by 0.50.

Now there are two additional complications that need to be considered. First, the males do not breed just once but do so two to five times in their lifetimes. Second, inexperienced, first-time breeders are less efficient at rearing chicks than experienced males. Fortunately, the actual productivities of experienced and inexperienced males were known. The final calculation showed that if the males sharing wives are brothers and if each male breeds only twice in his lifetime, then, altruism is likely to evolve. One fourth of the benefit is in fact greater than half the cost. But if males breed for five seasons, one fourth of the benefit is less than half the cost, and altruism is not predicted. It may seem a bit surprising that when males breed for five seasons, altruism is not predicted, but it is predicted when they breed only for two seasons. The answer to this apparent riddle is simple. Since experienced breeders do rather well without any additional help, males breeding for five seasons will lose more than they will gain from the additional male. But if a male breeds only twice, one of those attempts will be as an inexperienced individual and that is when he will gain from the additional male's efforts. Maynard Smith and Ridpath did not know whether a typical male breeds twice or five times. Nor did they know if males sharing wives were brothers. Nevertheless, this example shows how the inclusive fitness theory and Hamilton's rule permit us to make precise calculations and clear predictions about how animals will behave.

Why Do Lions Live in Groups?

I began this book by a consideration of the gregarious habits of the lion. George Schaller's famous work on the lions of the Serengeti National Park in Tanzania has since been continued and refined with more sophisticated techniques by Craig Packer and his colleagues. Packer's painstaking, long-term observational approach in the Schaller style combined with the power of molecular technology has yielded penetrating insights into the costs and benefits of lion social life. Recall that lion prides consist of several (usually two–nine) adult females and their dependent young and some (usually two–six) adult males. Since females do most of the hunting, let us first consider their point of view, by which of course we mean how natural selection might act on the females. There is good data on hunting success as a function of group size. When food is plentiful, group size has no effect on hunting success, but when prey are scarce there is a clear effect of group size on hunting success. During prey scarcity, hunting success, measured as kilograms of meat procured per female per day, is highest for lone females and for groups of five to six females, whereas it is significantly lower for groups of intermediate size. However, lionesses in prides containing less than five females hunt in as large a group as possible (instead of hunting solitarily) and lionesses living in large prides hunt in smaller groups of four to five individuals. Certainly the smaller prides and also to some extent the larger prides seem to be sacrificing hunting efficiency by foraging in suboptimal group sizes in spite of having the opportunity to achieve one of the two optimal group sizes.

Group sizes are therefore not determined merely by foraging efficiency. An important additional advantage of group living comes from the ability of all the mothers in a group to pool their offspring in a communal nursery and thus protect them, especially from infanticidal males. The lionesses appear to live in larger than optimal groups and sacrifice foraging efficiency in order to better protect their offspring and gain other long-term advantages such as superior defense of their territory. But there is no altruism here, because although lionesses in the observed group sizes hunt suboptimally, their reproductive success (or fitness) is highest in the most commonly observed group sizes. Although all females in a pride are closely related, there appears to be no opportunity or need for nepotism; mutual benefit is sufficient to explain the observed phenomena. The same cannot be said of the males.

Why do males tolerate each other in the pride? Indeed, not only do they tolerate each other but they cooperate in patrolling the pride's territory and in chasing away foreign males who would otherwise copulate with the pride's females. The addition of each extra male to the coalition of males in a pride leads, on average, to the production of an additional 0.64 surviving offspring. This was thought earlier to be sufficient reason for the males to cooperate because it was mistakenly believed that all males in a pride father about an equal proportion of offspring regardless of the number of males in the pride. Careful paternity analysis using highly variable (and therefore highly informative) DNA markers by Packer and his colleagues has now revealed that while reproduction is almost evenly shared in small groups it is rather unevenly shared in large groups. But while the males forming a coalition in small groups tend to be unrelated, those in large groups are often close relatives. This correlation of low relatedness with even sharing of reproduction, on the one hand, and high relatedness with uneven sharing of reproduction, on the other, demonstrates the importance of relatedness or nepotism in the evolution of cooperation. In short, lions seem willing to act as nonreproductive helpers in coalitions consisting of close relatives but are unwilling to do so in coalitions consisting of unrelated individuals, much as we would expect from Hamilton's rule, although the costs and benefits have not been measured and quantified in this case.

Why Are Honey Bee Workers Sterile?

Inclusive fitness theory is very powerful, and stated in the form of Hamilton's rule it permits us, as we have seen repeatedly, to explain many apparent paradoxes in animal behavior. The theory does, however, seem to suggest that altruism should be a rare phenomenon. Hamilton's rule states that the benefit (relatives reared) multiplied by their relatedness to the altruist should be greater than the cost (usually offspring lost) multiplied by their relatedness to the altruist. Usually there are no genetic relatives more closely related than offspring. Our closest genetic relatives are our offspring and our siblings, both of whom are related to us by 0.50. If an altruist gives up offspring and raises relatives, then he must somehow raise more relatives than the number of offspring given up. Only then will altruism be favored. Giving up one child should lead to the survival of at least two siblings; otherwise there is no particular advantage in giving up offspring in favor of siblings. If altruists are to be fitter than selfish individuals, then they should work harder than selfish individuals because they are giving up offspring for equally related or more distantly related relatives. This should make the evolution of altruism difficult because it requires the

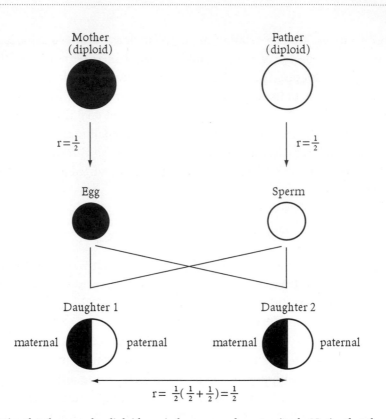

Genetic relatedness under diploidy, as in humans and most animals. Notice that the relatedness *(r)* between two full sisters is 0.50.

additional condition that altruists be better at raising young than selfish individuals.

Altruism, especially the kind where an individual becomes sterile and completely gives up offspring in favor of rearing relatives, is indeed rare in the animal kingdom, with the conspicuous exception of social insects such as ants, bees, and wasps. Most ants and many bees and wasps live in colonies, such as that of the honey bee, where sterile female workers rear the offspring of their queen. Why should altruism be so common in these insects? It must have been a tremendously exhilarating experience for Hamilton when he realized that it is only in the ants, bees, and wasps that there is indeed a class of relatives who are more closely related to each other than parents are to offspring.

Ants, bees, and wasps belong to an insect group called Hymenoptera. Among the many peculiarities of the Hymenoptera, the one that must rank above all is that they can reproduce parthenogenetically as well as sexually. Fertile females, the queens, can mate and store sperm received from their mates in their spermathecae. The mated females of course go to great lengths to keep the sperm alive and healthy. They have accessory glands that supply nutrients to the sperm and keep them viable. This ability is very useful to the queens, who make a risky mating flight only once in their lifetime. Once they become full-fledged queens, they usually never leave the nest and are heavily guarded and protected by the workers. The queen is just too precious to be allowed to leave the nest for mating or for anything else. The altruistic workers have invested everything in the welfare of their queen. Should the queen die, the workers will have no future and they will have lost their only instrument of gaining fitness. (Workers in many species can lay a very small number of unfertilized eggs upon the death of the queen, but the fitness so gained is almost negligible.)

The hymenopteran queen is perfectly capable of choosing the sex of her offspring. When she wants to produce daughters, she lets some sperm flow from the spermatheca into the oviduct so that fertilized eggs are laid. When she wants to produce sons, she closes the spermathecal duct and prevents sperm from reaching the oviduct so that only unfertilized eggs are laid. The fertilized eggs develop into females, while the unfertilized eggs develop into males. (Diploid males will develop on rare occasions from fertilized eggs but we can ignore that possibility for the purpose of our discussion.) This has great adaptive significance for the queen because it permits her to produce only daughters who can act as her workers when the colony is young and in need of a strong worker force; sons are useless in this regard because they never work.

This ability of the queen has another interesting consequence. Since the males develop from unfertilized eggs, they are haploid, having only one set of chromosomes (or genes, since chromosomes are nothing but strings of genes), which they receive from their mother through the egg. They have no paternal chromosomes because they have no father and thus could not receive paternal chromosome through sperm. But the females, whether queens or workers, always develop from fertilized eggs and are diploid, having the usual two sets of chromosomes, a maternal set transmitted through the egg and a paternal set transmitted through the sperm. Since the males are haploid and the females are diploid, the Hymenoptera are haplodiploid. In almost all other insects and other animals, both males and females develop from fertilized eggs and are both diploid. All diploid organisms produce gametes—sex cells, either eggs or sperm—by a process of division called meiosis. Each egg or sperm receives only half the number of chromosomes of the parent; otherwise the number of chromosomes would go on doubling when egg and sperm fuse. When contributing half the number of chromosomes to each egg or sperm, the diploid parent does not keep its paternally and maternally derived chromosomes separate. Instead, the paternally and maternally derived chromosomes are shuffled and randomly assigned to each gamete. The human male, for example, has 46 chromosomes, 23 received from his father and 23 from his mother. When sperm are made through meiosis, each sperm receives only 23 chromosomes from this set of 46. But of the 23 chromosomes in each sperm, each chromosome has an equal chance of being the man's father's or his mother's. Thus the sperm produced by a single male are genetically diverse and so are the eggs produced by each individual female in any diploid species. This is what makes brothers and sisters quite different from each other; they share only 50 percent of their genes with each other.

The hymenopteran male, however, has no paternal chromosomes to shuffle with the maternal ones. He only has the maternal set, and so he sends a complete maternal set to each sperm. All the sperm produced by a single male are thus genetic clones of each other. This means that two daughters having the same father and mother are no longer related to each other by just 0.50, as they are in diploid organisms. In the Hymenoptera, daughters share 50 percent of their maternal genes with each other because the eggs were produced by meiosis, and also share 100 percent of their paternal genes with each other because the sperm are all clones. Consequently, two daughters with the same father and mother are related to each other by 0.75 (this is the average of 50 percent genes shared between the eggs and 100 percent genes shared between the sperm). A hymenopteran worker who gives up offspring (related to her by the usual 0.50) in favor of sisters (related to her by 0.75) has a special advantage. Since she is more related to her sisters than she would be to her own offspring, she gets more inclusive fitness by raising one sister than by raising one son or daughter. Thus she actually has to work less as an altruist than she has to as a selfish individual. Stated in another way, she gets more fitness for the same amount of work as an altruist than she does as a selfish individual. For this reason, altruism can evolve more easily in the Hymenoptera than it can in diploid organisms.

Altruistic sterility of the kind exhibited by honey bee workers appears to have evolved a dozen times during the course of evolution in the Hymenoptera alone. In all the other animal groups, which of course outnumber the Hymenoptera by orders of magnitude, altruistic sterility appears to have evolved only three or four times. This provides strong

tunnels in Africa, aphids, and some beetles—both males and females act as sterile workers.

Bert Hölldobler and E. O. Wilson tested this hypothesis of male-female difference in the propensity for altruism toward colony members in a novel way. They studied the African weaver ant *Oecophylla longinoda,* a close relative of the common Asian weaver ant *Oecophylla smaragdina.* Weaver ants construct large conspicuous nests on different species of trees by weaving together several leaves with silk. The rather long-legged reddish-brown workers are notable for their painful bite, which includes an injection of formic acid. Weaver ants live in very large colonies consisting of a single queen and two kinds of workers, the larger *major* workers who forage, build the nest, and take care of the queen, and the smaller *minor* workers who care for the eggs and young larvae. The adult workers cannot produce silk. The silk is produced by the larvae. In most insect species the larvae use their silk to spin cocoons inside which the pupae undergo metamorphosis. The silk that we use for making clothes we steal from the silk worm, which produces it to spin its cocoon.

The major weaver ant workers who need silk to hold leaves together during nest construction do exactly the same thing—they steal silk from the larvae. While some major workers maneuver and hold leaves together, other major workers hold partially grown larvae in their mandibles and weave them across the leaf seams. This makes the larvae release strands of silk from glands underneath their mouths. Silk is used by the larvae of other insects purely for the selfish purpose of spinning a cocoon for themselves. But the weaver ant larvae use their silk in an entirely altruistic way and do not make cocoons for themselves. Of course silk is metabolically expensive to make and a cheater larva that donates no silk or less silk than another larva can use all its resources for its own growth. Hölldobler and Wilson found that male larvae have

Cooperative nest building in the weaver ant *Oecophylla longinoda*. The nest is constructed of living leaves and stems bound together with larval silk. Some of the walls and galleries are made entirely of silk. *(Reprinted with the permission of Harvard University Press from Hölldobler and Wilson 1990.)*

smaller silk glands and contribute substantially less silk for nest construction than do female larvae, exactly as predicted by inclusive fitness
theory. Male larvae do, however, contribute some silk because the nest
protects them also and they care about themselves. But the female larvae
contribute much more (about ten times more) silk because they not
only care about themselves but care more for the rest of the colony than
the males do. Or, to put it more technically, the optimum amount of silk
donated for nest building is higher for the female larvae than for the
male larvae because of the asymmetries in genetic relatedness that we
have been discussing.

Can Animals Recognize Their Relatives?

We have seen several examples of nepotistic behavior in animals ranging from slime molds to insects, birds, and mammals. In order to be
nepotistic, animals must have a way of distinguishing between close
relatives and distant relatives. It is a curious fact that although Hamilton's inclusive fitness theory was put forward in 1964, no serious effort
was made to test the kin-recognition abilities of animals until almost 15
years later. In 1979, Les Greenberg published the results of an experiment that has driven behavioral biologists into a frenzy of activity.
Greenberg was a student in the laboratory of Charles Michener at the
University of Kansas, where like most of Michener's students at that
time, he was studying a little bee known as *Lasioglossum zephyrum*. This
bee lives in a maze of underground tunnels in the soil. Michener and
his students learned to rear the bees in the laboratory by providing them
with soil enclosed by two glass plates. The bees take to this artificial
habitat quite readily. Another advantage these bees offer the experi-
menter is that they readily mate in the laboratory, so that Greenberg had
a number of bee stocks whose genetic relationships were all known to

him. It is typical for one of the female bees in a nest to sit near the entrance, guarding the nest and warding off intruder bees and other insects. Greenberg put the guard's efficiency to a severe test. He experimentally introduced intruders of the same species of bees from his laboratory stocks and measured the guard's responses. Greenberg presented guard bees with intruders who were the guard's sisters, aunts, nieces, cousins, or were unrelated individuals. He obtained the remarkable result that the probability that the guard would let the intruder pass her and enter the nest was tightly correlated with the genetic relationship between the guard and the intruder. Closely related intruders were more likely to be accepted while less related or unrelated intruders were more likely to be rejected. This suggests that the guard bee not only can discriminate between relatives and nonrelatives but also can tell who is more related and who is less related.

After Greenberg published his findings, behavioral biologists rushed to test the possible kin-recognition abilities of any animal they could get hold of. For the next 15 years, kin recognition became one of the most fashionable research areas in studies of animal behavior. Kin-recognition abilities of one sort or another have now been documented in marine invertebrates, mites, sweat bees, honey bees, several species of ants and wasps, termites, fishes, frogs and toads, iguanas, several species of birds, and a variety of mammals. In most cases, kin recognition is achieved by smell or body odor. We might think of every individual carrying on its body a relatively distinct odor label and in its brain an odor template, so that it can smell any encountered animal and match the label on the encountered animal's body with the template in its own brain and decide how closely related the encountered animal is to itself. In principle it is possible for every individual to have its own distinct label and distinct template so that every individual has a unique identity. If the labels and templates are genetically determined, then they will

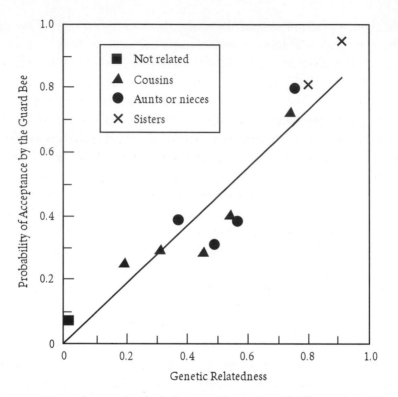

Kin recognition in the sweat bee *Lasioglossum zephyrum*. In artificially constituted laboratory colonies guard bees were presented with intruder bees whom they had never encountered before. The probability of acceptance into the nest of the intruder bee by the guard bee was significantly positively correlated with the average genetic relatedness between the guard and the intruder. *(Reproduced with permission from L. Greenberg, "Genetic Component of Bee Odor in Kin Recognition," Science, 206 [1979], 1096; copyright 1979 by the American Association for the Advancement of Science.)*

vary between related individuals in a graded manner so that identical twins have identical labels and templates, siblings have slightly different labels and templates, nieces and nephews have more different ones, and cousins still more different ones, and so on. We already know of a situation where something very similar happens. This is the so called major histocompatibility system of the mammals, which makes it possible for our bodies to recognize and reject any foreign tissue. We are unable to reject tissue from our identical twins; we react more strongly to foreign tissue from distantly related donors than we do to tissue transplanted from close relatives. It is the discovery of this and of ways to suppress the recognition system temporarily that made organ transplantation possible.

In most cases, however, the labels and templates used in kin recognition do not appear to be strictly genetically determined. Labels and templates are at least partially based on what animals learn about odors and on odors they acquire from the environment. There has been a great deal of interest in determining whether honey bees and other social insects can distinguish closely and distantly related individuals from within their own colonies. For instance, honey bee queens are known to mate with several males, store sperm from all of them, and use such mixed sperm to produce several different patrilines of daughters at any given time. Daughters belonging to different patrilines are half sisters. Inclusive fitness theory predicts that honey bee workers will give better and preferential care to their full-sister larvae (with whom they share 75 percent of their genes) than to their half-sister larvae (with whom they share only 25 percent of their genes). That would be the height of nepotism. Most of the larvae that worker bees rear will develop into future workers, but some of them will develop into future queens. Obviously, nepotism in choosing full-sister larvae over half-sister larvae for preferential treatment would be far more important in rearing larvae destined to be queens than in rearing larvae destined to be workers; it's the queens who are going to directly transmit genes to future generations.

The current state of our understanding of whether worker honey bees are so nepotistic as to prefer their full-sister larvae over their half-sister larvae for queen rearing can best be summed up in the words of Kirk Visscher, a prominent researcher in this field. In reporting his findings, Visscher writes, "Motive is established by genetic theory: it is clearly advantageous to a bee to invest her reproductive effort in a larva three times as closely related to herself as alternative larvae. Opportunity is established by our knowledge of the reproductive biology of honey bees: brood of varying relatedness to a given worker is always present in a colony. Means are established by the experiments reported here: bees can tell the difference between related and unrelated larvae, and differential treatment of larvae will influence which become queens. Like all circumstantial evidence, these observations must fall short of convicting *Apis mellifera* of nepotism, and it remains to be demonstrated that half-sister versus full-sister discrimination is actually an important element in queen rearing by colonies under natural swarming conditions, but given the strength of the case, it would be surprising if it were not."

7

A Primitive Wasp Society

Ropalidia marginata and *Ropalidia cyathiformis,* two species of wasps widely distributed in southern India, eminently qualify for the term primitively social. These wasps are called paper wasps because they build their nest from paper which they themselves manufacture from cellulose fibers scraped from plants. The nest is like a honey comb in that it has hexagonal cells, but it is much smaller than a bee's nest, rarely exceeding 500 cells, and much flatter, almost two-dimensional, so that except for the brood, the wasps are really on the nest and not in it. The number of wasps in a colony is also much smaller, rarely exceeding 100, than the number of bees in a colony. All this makes it easy for me to mark every individual wasp and make detailed observations on its behavior, its interactions with other members of the colony, and its contribution to the welfare of the colony. There are many interesting differences between these wasps and members of advanced insect societies such as ants and honey bees. Unlike colonies of ants and honey bees, these wasp colonies do not have a well-differentiated queen. The wasps in a colony all look alike. But only one individual in *R. marginata* and a small number of individuals in *R. cyathiformis* function as queens at any

A nest of *Ropalidia marginata*. *(Photo: R. Gadagkar.)*

given time. The wasps in a colony fight, and the winner usually becomes the next queen, but only for a while, because she is often challenged and driven away by one of the others, who then becomes the next queen, and so on. The individuals who are not queens at any given time act as workers—they do not reproduce but instead build the nest, forage for

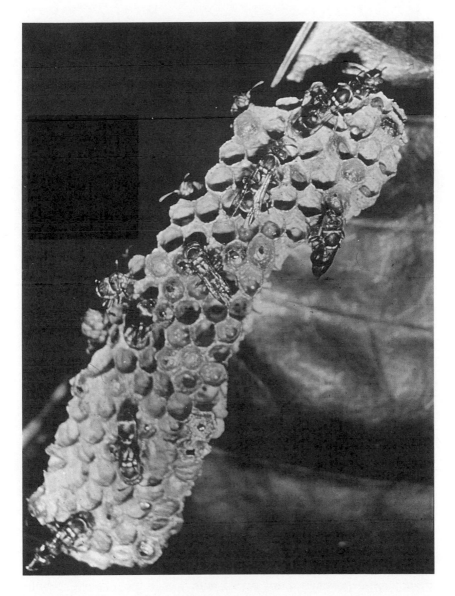

A nest of *Ropalidia cyathiformis. (Photo: R. Gadagkar.)*

food, and care for the brood. I mark the wasps with spots of quick-dry-ing paint of different colors so that I can differentiate each wasp from the others. In many cases I simply refer to a wasp by the color of the paint I have given—Red, Orange, or Blue.

New nests are started by one female or a group of females. If it is a single-foundress colony, the foundress acts both as queen and worker and manages all by herself to bring her eggs to adulthood. In a multiple-foundress nest, one of the foundresses assumes the role of queen while the others assume the roles of workers. The queen lays eggs in the cells of the nest, and when the eggs hatch into larvae they are fed on a diet of spiders, hemipteran bugs, and caterpillars and occasionally some nectar, by the queen herself in single-foundress nests and by workers in multi-ple-foundress nests. Not all workers work to the same extent or do exactly the same things, although they are quite flexible about what they will do in an emergency. Under normal conditions, some of the workers take on most of the burden of going out of the colony in search of food and building material. These are the foragers. Others specialize in stay-ing home and working on the nest and on the brood. Even among these, some are more aggressive toward other members of the colony, and we naturally call these the fighters. The remaining wasps also work on the nest but they are relatively quiet and spend much time just sitting and grooming themselves; these are the sitters. As the larvae complete devel-opment, they pupate in the same cell and undergo metamorphosis. The entire process of maturing from an egg into an adult wasp may take about 2 months. If the wasp emerging from the pupa is a *R. marginata* male, he will stay on the nest for about a week and then leave to lead a nomadic life, mate with some foraging female wasp, and die. In *R. cyathiformis*, the males spend their whole lives in the colony except for brief periods when they leave the nest, apparently to mate with wasps from other colonies. Mating never takes place on the nest. In neither species do the males take part in any aspect of social life; they do not forage, feed larvae, or build. The wasp society, like all bee and ant societies, is a female society—a matriarchy.

If the emerging wasp is a female, she appears to have a number of options open to her. She may leave to start a new nest all by herself, she may leave with a group of females to do so, or she may join females from other colonies to start a new nest. Alternatively, she may remain on the nest and assume the role of a worker in the colony of her birth. Or she

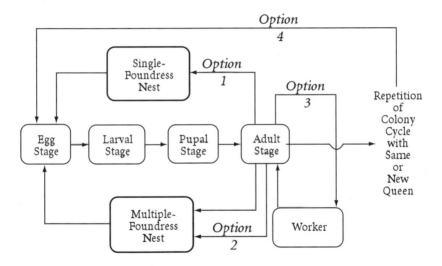

The perennial, indeterminate nesting cycle of *Ropalidia marginata*. Female wasps have at least four different options: (1) leave their natal nests to initiate single-foundress nests; (2) leave in a group to initiate multiple-foundress nests where they may become queens or workers; (3) stay on their natal nests as workers; and (4) stay on their natal nests to eventually take over as new queens. For schematic convenience, the egg, larval, and pupal stages are shown as distinct. In reality, there is considerable overlap between them. Similarly, a change of queens can occur at any time of the colony cycle. Also, new colonies may be initiated at any time of the year and may be abandoned at any time of the year and at any stage in the colony cycle.

may remain on the nest, work for some time, and eventually drive away the queen and take charge as the next queen in the colony of her birth. Of course such a power struggle may also take place between the co-foundresses in a new colony, so that one foundress may replace another even before producing any offspring.

Now why do we call this a primitive society? The social insects have evolved from solitary ancestors. In the transition from solitary to social life, it is reasonable to think that species would have gone through different stages of (1) nesting together without much interaction, (2) nesting together with interaction and some division of labor, and finally (3) nesting together with one morphologically specialized queen or a small number of such queens, who alone can reproduce. In *R. marginata* and *R. cyathiformis,* the wasps nest together, cooperate in nest building and brood care, and show some division of labor. But there is no morphologically differentiated queen incapable of performing the role of worker, and most workers can become queens if the opportunity presents itself. Besides, almost any wasp can start a nest and bring up her offspring by herself without participating in social life. Compared with the ants and honey bees, *R. marginata* and *R. cyathiformis* live in a less advanced or more primitive insect society.

Why Are Ropalidia Workers Altruistic?

In the advanced ant and honey bee societies, workers usually do not have any option but to work. They cannot leave to start their own nest and cannot usually mate and reproduce like the queen. We might say that they behave altruistically only because they have no choice. But in *R. marginata* and *R. cyathiformis,* most workers do have a choice. As we have seen, they can leave to start their own colonies as well as mate and reproduce in the same colony by driving away the queen. To drive away the queen and take her place may not be easy for all wasps, though they are probably plotting to do that all the time. But why do they not leave to start their own single-foundress nests where they can rear their own offspring, instead of staying to help the queen rear her offspring? Why do they behave altruistically even when they have the choice of being selfish? My first guess of course was that they gain more inclusive fitness as workers than as solitary nest foundresses because, being hymenopterans, they are more closely related to their sisters than to their own offspring. But do they usually rear full sisters in the nests where they function as workers or do they have to raise many half-sisters and other more distant relatives?

The Mating Habits of the Queen

My colleagues and I therefore set out to determine genetic relatedness between the members of wasp colonies. First we asked if the queens mate with only one male or with many males, like honey bees. To answer this question we studied an enzyme called esterase. Enzymes are proteins, which are made up of strings of amino acids. The net electrical charge on a protein molecule is a sensitive function of its amino acid sequence. The sequence of amino acids is determined by the sequence of the corresponding DNA in the genes. Many genes exist in multiple forms in a population and these forms lead to slight differences in the amino acid sequences of the proteins. If the protein is an enzyme, it can be separated from all other proteins and visualized through the use of dyes. The separation is done on a matrix consisting of starch or some other relatively inert material. We ground up individual wasps, applied the resulting juice to a gel, and subjected it to an electric field. The distance to which the esterase molecule moves on the gel is dependent on its charge and hence on its amino acid sequence. Thus one form of the gene may code for an esterase molecule that may be called *fast* and another form may code for a molecule that may be called *slow*. We determined the kind of esterase molecules present in the mother and her daughters in colonies of *R. marginata*.

In one experiment, for example, the mother and six of her daughters had only the fast form of esterase while four of her daughters had both fast and slow forms of esterase. Each individual has two copies of each gene, and the mother and six of her daughters were therefore homozygous—had two copies of the same fast gene each. Since each daughter gets one copy of each gene from her mother and one copy from her father, the father of these six daughters must have also carried the fast gene. But the remaining four daughters who had both fast and slow molecules must have had one copy each of the fast and slow genes. The mother could only have given them the fast gene and their father must therefore have carried the slow gene. Since males in the Hymenoptera are haploid, each male can have only one kind of gene, fast or slow. Hence the mother must have mated with at least two males, one carrying the fast gene and another carrying the slow gene. In this manner we discovered that *R. marginata* queens mate with a minimum of one to three males and simultaneously use sperm from different males and produce, like the honey bees, a mixture of full and half sisters among their daughters. We calculated that the average relatedness between sisters in a colony is 0.53. That is not much more than the 0.50 with which the worker is related to her own offspring. Thus workers are not much better off rearing sisters than their own offspring; and if they rear a mixture of brothers and sisters, they are distinctly worse off than they would be in rearing their own offspring.

Royal Pedigrees

There is a second reason that the genetic relatedness between the workers and the brood they rear may be lower than the theoretically expected values. Since some wasps can work for a while and then drive away the queen and take her position, the wasps in a colony may also be offspring of different mothers, in addition to being offspring of different fathers. We made a map of the nest about every other day and recorded the presence of eggs, larvae, and pupae in each cell. Since these do not move about, we could distinguish each egg, larva, and pupa in a cell from the

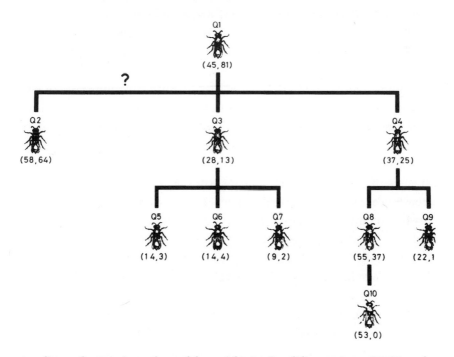

A pedigree of queens in a colony of the social wasp *Ropalidia marginata*. Q2, Q3, and Q4 were daughters of Q1; Q5, Q6, and Q7 were daughters of Q3, and so on. The question mark indicates that the relationship of Q2 to Q1 was unknown because both were already present at the beginning of the study. Of the two numbers in parentheses the first indicates the tenure in days of each queen and the second indicates the number of offspring she produced during her tenure.

eggs, larvae, and pupae in other cells. As soon as an adult emerged from its pupal case we marked it with a unique spot of colored paint. Thus all eggs, larvae, pupae, and adults were individually distinguishable. Since we always knew who the queen was at any given time and since there was only one queen at any given time, we knew the mothers of all eggs, larvae, pupae, and adults. In other words, we knew the genetic relationships of all members of the colony except for a very small number of individuals who were already present when we began our study.

With this information we constructed perhaps the first royal pedigrees ever worked out for an insect. The pedigrees from four colonies tell us that new queens may be daughters, sisters, nieces, or cousins of their immediate predecessor queens. Because at least some workers live long enough to rear the offspring of queens that take over the colony long after the death of their own mother, the relationship between workers and the brood they rear was even more diverse. We found the brood to be the workers' sisters, brothers, nieces, nephews, cousins, cousin's offspring, mother's cousins, mother's cousin's offspring, and even mother's cousin's grand-offspring. The workers therefore do not just rear their brothers and sisters, as was thought earlier. We calculated that, on the average, workers rear brood related to them by values ranging from 0.20 to 0.38. Now this is a far cry from the value of 0.50 that they would get if they left to start their own single-foundress colonies and rear offspring. So the plot thickens—why do the workers stay on the nest and altruistically rear somebody else's brood rather than acting selfishly and leaving to rear their own brood in their own nest?

The Nonnepotistic Advantages of Helping

Whatever advantage the workers get by their altruism, it can only be described as a nonnepotistic advantage because they are forgoing the opportunity to rear more closely related offspring and instead rearing less closely related relatives. One point perhaps needs clarification here. It must still be true that the inclusive fitness of workers is higher than that of solitary nest foundresses—otherwise even the modified form of Darwinian theory would not favor the worker strategy. But the reason for the greater inclusive fitness of the workers cannot be that they are more related to the brood they rear than is the case with solitary nest foundresses. Instead, the cost-benefit balance must be such that the workers get more inclusive fitness than solitary nest foundresses in spite of the fact that they rear brood less related to themselves than do solitary nest foundresses. Now why should the cost of altruism for these workers be so low, compared to the benefit accruing to the queen and her offspring, that altruism is favored by natural selection in spite of such low genetic relatedness between the altruist and the recipients of that altruism? I don't think I have the final solution to the paradox of worker altruism in *R. marginata*, but I have been pursuing three different leads.

Are Workers Inferior to Solitary Nest Foundresses?

One possibility is that all wasps are not born equal. Some may be more efficient at laying eggs, and these individuals may actually choose to become solitary foundresses; while others may be less efficient at producing eggs although they are perfectly capable of foraging and caring for the queen's brood, and these may choose to become workers. This can have a significant effect on our cost-benefit analysis. Thus the cost to an individual of becoming a worker would be the small number of eggs that she would have laid as a solitary nest foundress, and the benefit to the queen would be the large number of her offspring that the worker would rear.

How do we test this hypothesis? It is commonly believed that all wasps are born more or less equal and that suppression by dominant queens prevents workers from laying eggs. Immediately after their emergence from their pupal cases, we isolated a large number of wasps in individual cages and prevented them from being suppressed in any way by any other wasps. We provided these isolated, unsuppressed wasps with adequate food and building material and thus created suitable conditions for them to become solitary nest foundresses. Our reasoning was that if all wasps are born equal, they must all be able to initiate nests and start laying eggs. If all wasps are not born equal, the inferior ones should not be able to initiate nests and start laying eggs even though they have not been suppressed by any wasp after emerging from the pupal case. To our great surprise we found that only about half the wasps tested were able to initiate nests and lay eggs under these conditions. The other half died without doing so in spite of the fact that they lived long enough to have started laying eggs.

Thus we find that all wasps are not born equal—some are born inferior and are incapable of laying eggs even when they are not being suppressed by others. Even among those that built nests and laid eggs, some did so very soon after emerging while others took a long time to do so. We were able to classify wasps into early egg layers, late egg layers, and non–egg layers. The cost of giving up reproduction will naturally be very small for the non–egg layers, relatively small for the late egg layers, and very large for the early egg layers. Natural selection should favor the early egg layers to become solitary nest foundresses and should favor the late egg layers and especially the non–egg layers to become workers. The late egg layers have to wait a long time to develop their ovaries and then should survive until their offspring become independent of them. By contrast, wasps who choose to become workers can start working immediately because the queen will supply them with eggs. One of the reasons why some *R. marginata* workers choose to be altruistic and rear the queen's offspring rather than become solitary nest foundresses and rear their own offspring may be that they are not as good at producing eggs as others are. Notice that for this idea to work the workers need not necessarily be inferior to the solitary nest foundresses only in their egg-laying capacity; they could be inferior in other ways as well. As long as they are not as good at managing solitary nests as they are at working in someone else's nest, natural selection will favor such individuals to become workers. In other words, the cost of giving up reproduction is not so great for such individuals and therefore the benefit from altruism need not be so great either. Sociality will evolve in a species more easily if there are some individuals who are not very good at becoming solitary nest foundresses but are quite good at being workers in nests built by other individuals.

But why should some wasps be inferior in egg laying or anything else? To examine this question, we measured the rates at which larvae were fed in different nests and asked if better-fed larvae are more likely to become egg layers than poorly nourished ones. This is indeed the case. We found a statistically significant positive correlation between the amount of food an individual was given in natural colonies and its probability of becoming an egg layer in adulthood. Larval nutrition also influenced the time taken by the egg layers to start laying eggs—well-nourished larvae became early egg layers while poorly nourished larvae developed into late egg layers. It is not hard to imagine how the quality and quantity of nutrition received by larvae during early development can channel them into different developmental pathways leading to non–egg layers, late egg layers, and early egg layers. One can go a step further and ask why different larvae get different quantities of food. One possibility is that this is a result of parental manipulation: perhaps adult wasps "deliberately" undernourish (by which I mean that they are pro-grammed to do so and don't merely underfeed by accident) some larvae and give adequate nourishment to others so that some potential workers and some potential queens develop from the larvae. Indeed, it has been argued that such parental manipulation to ensure that at least some individuals become workers may be favored by natural selection under stringent ecological conditions, when the success rate of solitary nests can be very low. A mother who produces all selfish daughters who attempt to nest solitarily may have very few grandchildren, while a mother who manipulates a fraction of her offspring to become workers and help the remaining fraction of her offspring may leave behind more grandchildren, at least in some environments.

The other possibility, and a simpler one I believe, is that different larvae get somewhat different quantities of food merely by chance, without the adult wasps having to be programmed to feed larvae differentially. Such statistical fluctuations in larval nutrition can also lead to variability in the egg-laying and other abilities of the adult wasps. But of course we are now getting into the realm of proximate versus ultimate explanations. Whether the mechanism of feeding some larvae more and others less is based on the inefficiency of the adult wasps in rationing food equally or on genetic programming to feed differentially is a matter of the proximate mechanism of achieving differential larval nutrition. Of concern for us here is the ultimate evolutionary consequence of differential larval nutrition, and it seems reasonable that facilitation of the evolution of a worker strategy is one such consequence.

Is the Worker Strategy a Gamble?

Not all workers are doomed to die as workers. Some of them can replace the queen and take on her role. When several foundresses come together to start a multiple-foundress nest, those that become workers may not have joined as workers but may have come along hoping to become the queen. Are workers hopeful queens? Are they gambling on the hope of becoming a queen at some time before they die? Consider a simple illustration. Let us say a solitary foundress produces 10 offspring. But when two wasps get together they cooperatively produce not just 10 + 10 = 20 offspring, but, let us say, 21 offspring. It is true that all the 21 offspring will develop from eggs laid by only one of the two wasps. But who that lucky individual will be may be decided by chance. Figuratively speaking, the wasps may toss a coin and decide who will be the worker and who will be the queen. Each wasp then has a 50 percent chance of being the lucky winner, and on the average each wasp will be the queen half the time. An average gambler will produce half of 21 or 10.50 offspring, while an average solitary foundress will produce only 10 offspring. Naturally the gamblers do better than solitary foundresses and they will be favored by natural selection. Workers may therefore just be the losers in the gamble. And perhaps it is a worthwhile gamble. Many solitary foundresses we have seen are unsuccessful at producing even one adult offspring and those that are successful produce just one or two adult offspring. By contrast, an average queen of a multifemale nest produces about 78 adult offspring. Now even if there is only a small chance of becoming a queen, it might well be worth the gamble.

The Workers' Assured Returns on Their Investment

There is a curious advantage of living in groups, as workers do. A solitary foundress necessarily has to survive for the entire duration required for bringing up her brood. In *R. marginata* this takes 62 days. Even if the solitary foundress dies on the sixty-first day, she forfeits all her efforts because parasites and predators will take advantage of her absence from the nest and destroy her brood. A worker does not have such a serious problem. Even if she dies before bringing the brood under her care to independence, another worker from her colony will continue her job so that the brood will not die. If wasps always die before 62 days, say after 31 days, then solitary nest foundresses will never be able to rear any brood successfully, whereas workers can serially divide labor and each group of two workers can bring up one egg to adulthood. In other words, workers have relatively more assured returns on the investment of their labor while solitary nest foundresses do not. The extent of the disadvantage faced by solitary nest foundresses for this reason depends on what their probability is of dying before 62 days elapse.

Much like an actuary in a life insurance company, I have used the survivorship curve for a large population of wasps to estimate the probability that an average wasp will die before 62 days elapse. In the case of workers, I have given credit (fitness) to different workers depending on how long they live. This calculation shows that compared with workers solitary nest foundresses have a 3.60-fold disadvantage. Hence, all other things being equal, workers will break even with solitary foundresses even if they rear brood 3.60 times less related to them than solitary foundresses would do. Since solitary foundresses rear sons and daughters related to them by 0.50, workers can afford to rear brood related to them by as little as 0.50/3.60 = 0.14. We saw above that worker-brood relatedness values range from 0.20 to 0.38. This factor, which I have called assured fitness returns, is hence capable by itself of counteracting the problem of low relatedness seen in colonies of *R. marginata.*

8

Games Animals Play

The Prisoners' Dilemma

Two prisoners locked up in a cell face the following dilemma. Being sentenced for life, they have no hope of freedom. So they decide to dig a tunnel and escape. If they cooperate with each other and keep their activities secret, there is a small chance that they will both be able to escape. Let us call this the reward for mutual cooperation and give it an arbitrary value of 3 points. If prisoner A defects and reveals the escape plan to the jailer, he will be rewarded for his honesty and for foiling the escape plans of his partner and his sentence will probably be shortened. This is more valuable for him than the small chance of successfully escaping by mutual cooperation. Let us call this the temptation to defect and give it a score of 5 points (greater than the reward for mutual cooperation). At the same time prisoner B will have his sentence increased for plotting to escape. Let us call this the sucker's payoff and give it a score of 0 points. Of course, if prisoner B defects first then he gets 5 points and prisoner A gets 0 points. If both defect and give up plans to escape, then they continue with their sentence. Let us call this the

125

punishment for mutual defection and give it a score of 1 point (less than the reward for mutual cooperation but more than the sucker's payoff).

Unfortunately, neither prisoner can read the other's mind or predict what the other will do. Let us say prisoner B cooperates. Prisoner A will get the reward for mutual cooperation of 3 points if he also cooperates but he will get the temptation to defect of 5 points if he defects. So he is better off defecting. Alternatively, let us say prisoner B defects. Prisoner A will get the sucker's payoff of 0 points if he cooperates and the punishment for mutual defection of 1 point if he defects, so again he is better off defecting. Thus prisoner A should defect no matter what prisoner B is going to do. The same applies to prisoner B, who should also defect no matter what prisoner A is going to do. So both should defect and be content with the reward for mutual defection of 1 point each. However, if both had cooperated, they would each have gotten the

The prisoner's dilemma game. The payoff to player A is shown with illustrative numerical values.

		Player B	
		Cooperate	Defect
Player A	Cooperate	$R = 3$ Reward for mutual cooperation	$S = 0$ Sucker's payoff
	Defect	$T = 5$ Temptation to defect	$P = 1$ Punishment for mutual defection

Source: Reprinted with permission from R. Axelrod and W. D. Hamilton, "The Evolution of Cooperation," *Science*, 211 (1981): 1392; copyright 1981 by the American Association for the Advancement of Science.

reward for mutual cooperation of 3 points. Hence the prisoners' dilemma—to cooperate or to defect? Mutual cooperation is best for both, but since neither knows what the other will do, it is safer for each to defect.

Tit for Tat—an Unbeatable Strategy

So far we have considered a situation where the two prisoners meet and interact only once and never meet again. But think of it as a game that you and I play repeatedly and we accumulate the points we get, game after game. How do we get one up on each other and get more points than the other at the end of the day? Now there is some hope of solving the dilemma because we can begin to predict each other's psychology. If I have some inkling of whether you are going to cooperate or defect, then I might be better able to plan my strategy. There is no better way to appreciate the problem than to actually play a series of games with someone and award points as described above. Incidentally, such games are taken quite seriously by many people. The political scientist Robert Axelrod invited people from across the globe to submit strategies that they might adopt if they were caught in such a repeated prisoner's dilemma. Entries came from economists, psychologists, sociologists, political scientists, and mathematicians. Axelrod pitted the strategies he received against each other in a round robin tournament on his computer. To everybody's surprise, including that of Axelrod, the winner was a strategy called Tit for Tat submitted by Anatol Rapaport of the University of Toronto. Tit for Tat is a rather simple strategy. It always cooperates in the first move with any stranger, and thereafter it does exactly what the opponent did the last time around. There were many strategies which beat Tit for Tat, but when each strategy had to face all other strategies in the field, Tit for Tat emerged as the clear winner. Now

Axelrod made public this result of the superiority of Tit for Tat and again invited people to come up with winning strategies. Naturally, people specifically attempted to do better than Tit for Tat. Many strategies were submitted and several of them were attempted improvements over Tit for Tat. Anatol Rapaport had so much confidence in his Tit for Tat strategy that he submitted it again, unmodified. Guess what, Tit for Tat won again!

Next, Axelrod played his computer tournaments with an evolutionary twist. He let each strategy reproduce itself in proportion to its success in accumulating points. So there were many more players playing Tit for Tat because it kept winning. The end result of this was that Tit for Tat ended up as the most common strategy in the population. Thus Tit for Tat is a robust strategy.

Three properties appear to make Tit for Tat an unbeatable strategy in a round robin tournament with all possible strategies. Tit for Tat is nice, forgiving, and retaliatory. It is nice because it always cooperates in the first move. It is forgiving because it immediately responds by cooperating if the opponent cooperates even once after any number of previous defections. It is retaliatory because it responds by immediate defection as soon as the opponent defects even once after any number of previous cooperations. Robert Axelrod has written a book called *The Evolution of Co-operation* in which he explains how cooperation can indeed evolve and be stable in a population of egotists, be they microorganisms or warring nations. In his foreword to Axelrod's book Richard Dawkins says, "This is a book of optimism. But it is a believable, realistic optimism, more satisfying than the naive, pie-in-the-sky optimisms of Christianity, Islam or Marxism . . . As Darwinians we start pessimistically by assuming deep selfishness, pitiless indifference to suffering, ruthless heed to individual success. And yet, from such warped beginnings, something that is in effect, if not necessarily in intention, close to amicable brotherhood and sisterhood can come. This is the uplifting message of Robert Axelrod's remarkable book . . . The world's leaders should all be locked up with this book and not released until they have read it. This would be a pleasure to them and might save the rest of us."

Animals Caught in a Prisoner's Dilemma

Animals are often caught in a prisoner's dilemma. Consider the problem of aggression. How should an animal behave towards its opponent? Should it attack and play what we will call a Hawk strategy or should it be mild and play a Dove strategy? Let us borrow an example suggested by John Maynard Smith. If I play Hawk and win, I get 50 points. But if I play Hawk and lose, I suffer injury and get minus 100 points. If Hawks play Hawks each will lose half the time and win half the time and on average they will get a score of $[(50)+(-100)]/2 = -25$. If Hawks play Doves, Hawks will get 50 points but Doves will retreat and get 0 points (no injury). If Doves play Doves, the contest is not easily settled. So there is much waste of energy displaying each other's strengths. Once settled, the winner gets 50 points but loses 10 points as the cost of displaying. The loser gets 0 points and also pays 10 points as the cost of displaying. On average each Dove (given that each wins and loses half the time) gets $[(50-10)+(0-10)/2] = 15$ points. Now the dilemma is not so much for individual players as it is for evolution.

The population cannot end up with all Hawks because when everybody is a Hawk, Doves do better; a Hawk-Hawk encounter yields -25 while a Dove-Hawk encounter yields 0. But the population cannot end up with all Doves because in a population of pure Doves, Hawks do better; a Dove-Dove encounter yields $+15$ while a Hawk-Dove encoun-

The game between a Hawk and a Dove.

Payoffs: Winner   +50   Injury   −100
          Loser   0   Display   −10

Payoff Matrix: average payoffs in a fight to the attacker

		Opponent	
		Hawk	Dove
Attacker	Hawk	$(50+ (−100))/2 = −25$	+50
	Dove	0	$((50 − 10) + (−10))/2 = +15$

Source: Krebs and Davies 1993, after Maynard Smith 1976.
Notes:
1. When a Hawk meets a Hawk, we assume that on half of the occasions it wins and on half the occasions it suffers injury.
2. Hawks always beat Doves.
3. Doves always immediately retreat from Hawks.
4. When a Dove meets a Dove, we assume that there is always a display and the displayer wins on half the occasions.

ter yields +50. So the population will swing from Hawks to Doves and vice versa and will always consist of a mixture of Hawks and Doves.

But animals have various ways of getting out of such tricky situations. To illustrate one such way Maynard Smith has introduced the idea of a Bourgeois strategy. A Bourgeois is one who sticks to convention and plays Hawk when he is in his own territory and plays the Dove strategy when he is in the opponent's territory. If you work out the arithmetic, you will see that Hawks and Bourgeois can invade a pure population of Doves. Similarly, Doves and Bourgeois can invade a pure population of Hawks. However, neither Hawks nor Doves can invade a pure population of Bourgeois. Thus the Bourgeois strategy is termed an evolutionarily stable strategy.

A Hawk-Dove-Bourgeois game.

Payoffs: Winner +50   Injury   −100
          Loser   0   Display −10

Payoff Matrix: average payoffs in a fight to the attacker

		Opponent		
		Hawk	Dove	Bourgeois
	Hawk	−25	+50	+12.5
Attacker	Dove	0	+15	+7.5
	Bourgeois	−12.5	+32.5	+25

Source: Krebs and Davies 1993, after Maynard Smith 1976.
Notes:

1. When a Bourgeois meets either a Hawk or a Dove, we assume it is the owner of the territory half the time and therefore plays Hawk, and the intruder half the time and therefore plays Dove. Its payoffs are therefore the average of the two cells above it in the matrix.

2. When a Bourgeois meets a Bourgeois, on half the occasions it is the owner of the territory and wins, while on half the occasions it is the intruder and retreats. There is never any cost of display or injury.

Bourgeois Butterflies

A butterfly species known as the speckled wood is a good candidate for the label bourgeois. Speckled woods spend the night in the forest canopy. With the warmth of the morning sun, they become active, and the males come down to the forest floor and bask in little pools of sunlight created by the penetration of the sun's rays through the forest canopy. Not all males can find sunspots all the time, so some of the males remain in the forest canopy and don't come down until they locate a sunspot. The sunspots of course keep moving with the sun and the males occupying these sunspots also move along and stay warm. Nick Davies, who studied speckled woods in Wytham woods near Oxford, England, found that the males actually treat their individual sunspots as their territories and actively defend them against other males arriving on their sunspots. They do this because the females like to visit sunspots too and Davies showed that males occupying sunspots are more likely to have opportunities to mate than males remaining in the canopy. But if a second male lands on an occupied sunspot, the resident male flies toward it and engages it in an apparently harmless spiral flight where both males flutter close to each other in mid-air and bump into each other. But within a few seconds, one of them returns to the canopy and the other returns to his perch in the sunspot.

Davies marked several butterflies with a Magic Marker and discovered that in every case it is the original owner of the sunspot who wins

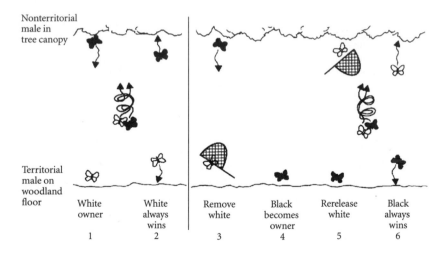

Speckled wood butterflies in sunspots: an experiment showing that the rule for settling contests for territories is the resident always wins. This experiment was done three times, each time in a different territory with a different pair of color-marked males. In the figure one male is represented as black and the other as white. *(Drawing: Tim Halliday; reproduced with permission from Davies 1978.)*

in the spiral flight and it is always the intruder who retreats. This was true even if the resident was an old and tattered weak-looking male. It was also true when the resident male had occupied the sunspot only for a few seconds and the intruder had previously occupied the same sunspot for much longer—the current resident always wins. Davies has described the spiral flights as short conventional displays where the owner says, "I was here first," and the intruder says, "Sorry, I didn't know there was anyone occupying the sunspot, I'll retreat to the canopy."

What if both males think that they are the true owners? This does not normally seem to happen but Davies made it happen on five occasions. He managed to introduce a second male into an occupied sunspot without the resident male noticing and apparently without the second male realizing that the sunspot was occupied. The first male to fly about was noticed by the other and then the two males, each thinking that he was the rightful owner and probably aghast at the other not making the usual civilized retreat, went through spiral flights that lasted ten times as long as the usual conventional flights. The conclusion is that the bourgeois butterflies normally follow a convention by which the intruders accept defeat immediately without a serious fight because contests may be costly and a sunspot as a territory is not very difficult to find.

Do Animals Play Tit for Tat?

Using a rather ingenious technique, Manfred Milinski has shown that sticklebacks employ Tit for Tat. The technique involves subjecting a pair of sticklebacks to a predator and examining how the two fish cooperate with each other in inspecting the predator. Inspecting the predator is important because gaining information about a potential predator is of great value in dealing with it. The goal of Milinski's research was to see which individual takes the lead in inspecting the predator and whether this depends on the participation of the other individual in the risk of predator inspection. But it was not easy for Milinski to force the sticklebacks to cooperate or not to cooperate. So he used only one individual, but he placed a mirror against the wall of the tank so that the stickleback thought that there was another individual in an adjacent area. Sometimes Milinski placed the mirror parallel to the wall of the tank, so that the stickleback thought that the other individual was moving in pace with it while it was moving toward the predator. At other times Milinski placed the mirror at an angle, so that the reflection moved more slowly and soon disappeared. The stickleback then got the impression that the other individual was lagging behind and waiting for it to go first to the danger spot. Milinski calls the parallel mirror the cooperating mirror and the mirror at an angle the defecting mirror. The readiness of the stickleback to approach the predator depended on whether the mirror was cooperating or defecting, and thus on whether the stickleback thought that the other individual was cooperating or defecting.

All three requirements of the Tit for Tat strategy appeared to be fulfilled. The sticklebacks were nice because they cooperated on the first move. The sticklebacks had to begin moving toward the predator before the reflection did. The sticklebacks were retaliatory because if the reflection did not move in pace, they did not approach the predator as often or as close as they did with the cooperating mirror. The sticklebacks were also forgiving because they would try repeatedly even when the reflection did not move in pace.

Are Animals Conscious of Their Actions?

Throughout this book I appear to attribute a certain level of intelligence and consciousness to animals. Is this fair or does this constitute a flaw in my arguments? This problem became really acute with the advent of sociobiology and the interpretation of animal behavior as nepotistic. Are animals being consciously nepotistic? For those who tend to draw moral lessons from animal behavior, it somehow seems worse if animals are consciously nepotistic rather than being blindly programmed. There is also the problem of whether or not animals, and especially insects, not to speak of slime molds, have sufficient intelligence to be able to be consciously nepotistic. When I was a student, I had a friend who studied elephant sociobiology, but he seemed to do so with a large dose of skepticism. He was very talented and had ingenious ways of expressing his subtle dissatisfaction with his research supervisor's scientific beliefs. He once showed me a cartoon that he had drawn to amuse himself. It pictured an elephant calf that had fallen into a pit and an adult elephant standing above it making calculations about its genetic relatedness to the calf, apparently unable to decide whether or not to rescue the calf.

The question of whether animals are consciously nepotistic seemed invidious, and most researchers in sociobiology were quick to point out that it is not necessary for animals to be conscious of their actions. The theoretical predictions about which behaviors will be favored and which disfavored by natural selection will hold perfectly well even if the animals are blindly programmed to do what happens to be evolutionarily wise. Indeed, the strength of the theory is that the same rules apply from slime molds to humans. But in our eagerness to prevent the problem of animal intelligence from becoming a stumbling block in the acceptance of our theories, I think we all went overboard and ignored the possibility of animal intelligence! After all, our theories will work equally well if animals consciously do what is evolutionarily wise. Consciousness and intelligence are not harmful to the theory; it is just that they are not essential to the theory, and if you don't believe in them there is no need to disbelieve the theory on that account. Now that sociobiological theories are more widely accepted, at least in the context of animals, and we have grown out of our initial nervousness, I think it is time we take a fresh, more relaxed look at whether animals do indeed behave intelligently. Do they think about what they are doing? Are they conscious of their actions? But such questions are seldom discussed by ethologists. There appears to be a strong taboo among ethologists against exploring the animal mind in this fashion.

What are the reasons for this fear of studying animal intelligence? First there is the question of definitions. How can we study animal intelligence, thinking, and consciousness, people argue, if we cannot even define these terms accurately? In some ways it is easier to define what is not intelligent behavior. The French naturalist Henri Fabre did a curious experiment with a digger wasp that constructs burrows in the ground to rear its brood. Having built a burrow, it hunts a cricket meant to serve as food for its larvae, places the paralyzed cricket a small distance from the burrow, enters the burrow to inspect it, and then returns to take the cricket in. When the wasp was inspecting the burrow, Fabre moved the cricket a little distance away from where the wasp had placed it. The wasp returned to take the cricket and discovered its absence. Finding the cricket soon enough, it once again placed it a small distance from the burrow, and went back to inspect. Fabre of course shifted the position of the cricket again but the wasp once again discovered the cricket, placed it near the burrow and went back to inspect. After 40 unsuccessful attempts to make the wasp take the cricket directly into the burrow without an intervening burrow inspection, Fabre gave up in exasperation. Obviously the wasp was incapable of realizing that since the burrow had been inspected many times in the recent past, the cricket could now be taken directly into it; or of realizing that since it was simply not succeeding in taking the cricket into the burrow, it should try a little variation in its sequence of behaviors. This machine-like, unintelligent behavior on the part of the digger wasp illustrates negatively what is meant by intelligence better than any definition of intelligence can. The important point is that an intelligent animal should be able to respond to variable and unexpected stimuli in a manner that is variable but appropriate to a given context.

Another familiar objection, especially to the possibility of intelligent behavior in animals as small as insects, is that small brains cannot possibly engender such behavior. There is little merit to this argument. It is being increasingly realized that it is not the size of the brain or the number of neurons in it but the quality and quantity of connections that really matter. The small size of insect brains should not therefore deter us from investigating possible examples of intelligent behavior on their part.

Donald Griffin, now at Harvard University, has achieved the status of a lone crusader in the cause of the study of animal intelligence and thought processes. His books provide a fascinating commentary on the complex and intelligent things that a wide variety of animals are capable of doing. I think his books provide an even more fascinating commentary on how most people, especially those who study animal behavior, have had a closed mind on the question of animal intelligence. The burden of Griffin's message is that viewing animals as being in a state comparable to human sleepwalkers will never let us find out if animals have conscious experiences; rather the question of animal consciousness should be treated as an open one. Griffin uses three categories of evidence in support of the claim that nonhuman animals may have conscious thoughts. First, he reviews a number of most fascinating examples of the versatile adaptability of animals to novel challenges—exactly the opposite of the behavior of Fabre's digger wasp. Second, he points out that detecting a neurophysiological correlate of conscious thinking is a definite possibility. This involves measuring a class of electrical impulses from the brain that are not direct responses to external stimuli but that apparently are affected by internal processes in the brain and hence appropriately called event-related potentials, EPR for short. There appears to be evidence that P300, an EPR that lasts for 300 milliseconds, may be correlated with thought processes in humans. The fact that P300s are seen in monkeys and other animals opens up the possibility of detecting thought processes in animals too, and efforts in this direction have already begun. Finally, and perhaps most important, Griffin argues that communicative behavior in animals provides an especially useful window on animal minds. We can only use this window effectively, however, if we stop thinking of animal communication signals as what Griffin calls "groans of pain" and start thinking of them as an attempt on the part of animals to assess other animals' moods and thoughts and predict their probable behavioral responses. The Dutch zoologist Frans de Waal has described some incredible instances of chimpanzee intelligence in his provocatively titled book *Chimpanzee Politics.* I will describe some of these in the next chapter and take courage from de Waal and describe some of my own observations on insects under what you will agree is an even more provocative title, "Wasp Politics."

9

The Fine Balance between Cooperation and Conflict

Domestic Conflicts in a Bird Family

Birds exhibit, more than any other group of higher animals, such "noble virtues" as monogamy, pair bonding for life, male parental care, and cooperative efforts by both parents in nest building and care of the chicks. Not surprisingly, these virtues of the birds are often extolled by poets and philosophers, especially when they are admonishing fellow humans. As we probe deeper into the secrets of bird family life, however, we find many unexpected domestic conflicts coming to the fore. A particularly startling revelation has come from the recent use of DNA technology to determine the parentage of chicks being reared in the nests of monogamously paired parents, much like the work of forensic laboratories in resolving cases of disputed parentage among humans. Many species that were fondly thought to be monogamous have turned out to be rather promiscuous. Females from apparently monogamous pairs often mate, on the sly as it were, with males from neighboring monogamous pairs and lay at least some eggs that are not sired by the partners who help them in parental duties.

A novel and more complicated domestic conflict has recently been documented by Norwegian scientists. T. Slagsvold, T. Amundsen, and S. Dale conducted a four-year study of the breeding biology of the blue tit, a small passerine bird, not unlike the common house sparrow. These birds are monogamous and both parents share in parental duties. The female lays about 10 eggs in a span of about 10 days and incubates them. The male does not help with the incubation, but he feeds the female while she incubates, and later, when the chicks hatch, both parents feed them. When should the female start incubating? If she starts too early (say, as soon as she lays her first egg), the chicks will hatch on different days and the parents will have a very asynchronous brood to take care of. If she starts late (say, after she has already laid all her eggs), the chicks will all hatch at almost the same time and the parents will have a very synchronous brood.

It turns out that synchronous and asynchronous broods have very different consequences for the male and female parents. This was discovered by artificially manipulating broods to produce especially synchronous or asynchronous broods. Males had a higher chance of surviving to breed again the following year if they cared for an asynchronous brood than if they cared for a synchronous brood. Conversely, females had a higher chance of surviving to breed the following year when they cared for synchronous brood rather than an asynchronous brood. Thus the mother is better off raising a synchronous batch of brood while the father is better off with an asynchronous batch of brood. The most likely reasons for these male-female differences are as follows. Males, while participating in parental care, are apparently not as conscientious as the females. They take care of the larger and stronger chicks and when these chicks are successfully fledged, they stop caring for chicks and concentrate on territorial defense and molting to enhance their future survival probabilities. The burden of difficult and prolonged care of late-developing, small, and weak chicks falls on the mother. When the chicks are all of more or less the same age, the mother thus has more help from the father, who in turn has to work harder because all the chicks satisfy his criteria of being big and strong. When the brood is asynchronous, however, the male benefits by stopping his work early while the female carries on alone, caring for the smaller and weaker chicks and in the process lowering her chances of being alive and fit to breed again the following year.

Now why should males and females be so different in their commitment to parental care? First, female parental care is more fundamental, and as soon as there is any opportunity for one of the parents to desert, it is usually the male who is the first one to seize it. This happens throughout the animal kingdom, and may be related to the fact that females invest more in their offspring, starting right from the substantial cost of an egg, while males invest much less, often nothing more than inexpensive sperm. Hence females have much more at stake in the survival of their offspring than males do. Second, the small, late-hatching chicks in a nest are more likely to be sired by neighboring males in extra-pair copulations, so that the resident male has even less interest in the welfare of these particular chicks. Thus one would expect a conflict between the two parents on the question of whether the brood should hatch synchronously or asynchronously. But this conflict remains hidden because only the female incubates, and thus only she can decide how synchronous the brood should be. In addition to the many examples of overt conflict seen throughout this book, there may be other such hidden conflicts that can be uncovered only by careful experimentation.

Until not too long ago, unexpected conflicts among animals were dismissed as being pathological. The evolutionary approach to animal behavior permits us to face such unexpected conflicts head on and even to predict when conflicts may occur and how they may be resolved. As a bonus, our understanding of animal behavior grows in richness. But if these revelations of domestic conflict in birds appear to make them unsuitable as models of good behavior, we must reflect on the fact that they are still able to maintain an external appearance of faithfully bonded monogamous pairs in spite of such hidden conflict.

Queen-Worker Conflict in Ants

We might argue that birds are not so socially evolved as some other species and hence they still experience a lot of conflict. What about the socially advanced ant societies, where the queen appears to be in complete control of the workers and the workers appear to have lost all options of revolting against the queen's authority? Is there still some conflict? It is true that many species of ants and bees have reached that pinnacle of social evolution where workers are locked into sociality and can neither lead a solitary life nor mate and reproduce—two prerequisites for revolting against the queen's authority. And yet if we look deeper, we see conflict here also. Even when workers cannot drive away the queen and take her role or leave the colony to start their own, natural selection would be expected to favor workers who get as much benefit as possible from the queen. Of course natural selection is impartial, so it would simultaneously act on queens to yield as little benefit as possible to the worker. Thus the conflict between queen and worker would come to the fore.

Consider an ant colony, where the workers are the queen's daughters. Because workers in the Hymenoptera are more closely related to their sisters than they would be to their own offspring, workers would be expected to cooperate with their queens in rearing the queen's female brood. Recall that workers are related to their brothers (the queen's male brood) by only 0.25. Thus workers should be more reluctant to rear their brothers and should prefer to rear their own sons. Rearing a combination of sisters and sons would be their ideal choice. The worker's sons are the queen's grandsons and are thus less related to her than her own sons would be. A queen would therefore prefer that workers rear her sons and daughters. Here is a region of conflict between queens and workers. This conflict can become intense because in many species of social Hymenoptera workers have not entirely lost their ovaries; they often have at least small ovaries and can lay a few unfertilized eggs, destined to be males. Queen-worker conflict over male production is now well known in many ant species. The workers attempt to lay haploid eggs and the queen attempts to eat them and then replace them with her own haploid eggs.

If the workers fail to win in this conflict by laying enough haploid eggs, all is not lost. It turns out that there is yet another weapon in their arsenal. After all, it is the workers who feed all the larvae and surely they can feed their sisters more than their brothers. In fact, considering that the workers are related to their sisters by 0.75 and to their brothers by 0.25, we should expect them to give three times as much food to their sisters as they would to their brothers. Shocking as it may seem, workers in many (but not all) ant colonies seem to do exactly this, although they may be somewhat imprecise in apportioning food in the ratio of 3 to 1. This is a rather striking confirmation of the theoretical expectation. But as they say, exceptions prove the rule. So we must find an exception to the rule that workers should feed their sisters three times as much as they feed their brothers and see if that exception is also found in nature.

Robert Trivers and Hope Hare, who originally made the bold suggestion that workers should bias their investment in the ratio of 3 to 1, have postulated two exceptions. In some ant species, several queens simultaneously lay eggs in each colony; these are called polygynous colonies.

Here the workers care for larvae that are not always their sisters because they may be the daughters of other queens in the same colony. The workers' relatedness to these larvae may be very low and would depend on the genetic relatedness between their mother and the mother of the larva concerned. Even if the mother of the larva was the sister of the worker's mother, the larva would be her cousin, and cousins are less closely related than sisters. So workers would not be selected to invest in female and male brood of the queens in the ratio of 3 to 1. In the few polygynous colonies studied from this point of view, the ratio of investment is not even approximately 3 to 1. Prediction confirmed once again.

The second exception that Trivers and Hare came up with is even more interesting. Some species of ants have abandoned the habit of producing a large number of sterile workers before producing future queens and males, since this is quite a costly undertaking. Instead they produce just enough workers to raid neighboring colonies of related species of ants and forcibly bring back worker pupae from the raided nest. These are called slave-making ants and the species providing the slaves (pupae), although none provides willingly, are called the slave species. When pupae of the slave species mature in their foster colonies, they wake up and start working; they don't seem to know that they have been kidnapped. But imagine what would happen if a mutation arose in the slave species that did not program them to invest in female and male larvae in the ratio of 3 to 1. Such a mutation does not suffer any great disadvantage compared with the wild type, because the ant slaves work for different species altogether and will yield them no fitness anyway. So natural selection should not be expected to have perfected the adaptation of the 3 to 1 investment ratio in slave species as effectively as it might have done in the nonslave species. This indeed appears to be the case.

If only for the sake of amusement, we can wonder who wins in each case of conflict. In normal monogynous colonies, the workers seem to have the last laugh because they are in charge of feeding and they can bias investment in male and female larvae in the ratio advantageous to them and not in the ratio advantageous to the queens. In polygynous colonies workers are forced to care for the brood of several queens and therefore cannot have their way; hence the queens benefit from their predicament. In slave-making species, queens benefit from the fact that the workers are aliens and have no interest in upstaging the queens. But the conflict is always there and it is often resolved in unexpected ways.

Worker-Worker Conflict in Honey Bees

Francis Ratnieks has come up with another twist to the story of conflict within the apparently harmonious colonies of advanced insect societies. Recall that if the mother queen mates with just a single male, the workers will all be full sisters and thus related to each other by 0.75 and to their brothers by 0.25. In such a situation, workers should prefer their own sons over their brothers. Ratnieks has argued that any worker should also prefer any other worker's sons (her nephews) over her brothers, because a worker is related to her nephews by half the value of her relatedness to her sisters and that comes to 0.375 when sisters are related by 0.75. Thus workers should have a common interest in revolting against the mother queen and laying their own male-destined eggs. The workers should not have much conflict among themselves because they would rather rear male eggs laid by each other than those laid by the queen.

But if the queen mates with several males and produces daughters by using sperm from different males, the workers will now quite often be stepsisters or half sisters, related to each other by only 0.25. Although each worker should continue to prefer to rear her own sons rather than her brothers, workers should now cease to prefer each other's sons. The son of a half sister would be related by only half of 0.25, which is 0.125. Each worker should now prefer the queen's sons over another worker's sons. Although their first preference would still be their own sons, they would not agree on which of them should produce the male eggs. Indeed, Ratnieks has argued that workers should police each other and

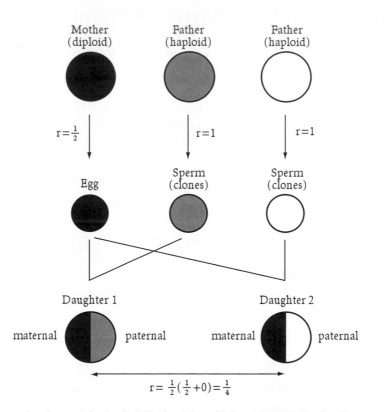

Genetic relatedness under haplodiploidy with multiple mating. Notice that the genetic relatedness between half sisters is 0.25.

destroy any eggs laid by each other because they would no longer (owing to multiple mating by the queen) prefer nephews over brothers.

The honey bee society is a good system to use to test this prediction since the queens are known to mate with 10 to 20 different males. Ratnieks collaborated with Kirk Visscher, the man whose reluctance to convict *Apis mellifera* of nepotism we saw earlier. Ratnieks and Visscher used the European honey bee *Apis mellifera* and asked if workers actually police each other by eating each other's eggs, as predicted by the theory. They found that while only 0.7 percent of the worker-laid eggs survived after 24 hours, 45.2 percent of queen-laid eggs did so after the same time period. It seems rather ironic that the queen ultimately benefits from the inability of the workers to agree on which one of them should lay the male eggs, although they all agree that it is not the queen but they who should be doing so. Is the queen's habit of mating with several males a strategy to disrupt the workers' unity and get them to fight with each other?

Disease As an Enhancer of Social Conflict

Parasites that cause diseases of varying intensities are ubiquitous in the natural world. But the role of disease in shaping the ecology and evolution of their hosts has only recently begun to be properly appreciated. Parasite load has, for example, been shown to be an important parameter that females use to assess the quality of their mates. In response, males are known to evolve elaborate secondary sexual characters to impress upon females their health in general and their ability to resist parasites in particular. Recently, an even more profound role that parasites play in modulating social evolution has come to light.

As we saw with the honey bees, queens in many social Hymenoptera mate with several males and simultaneously use sperm from different males to produce several patrilines of daughters. Daughters belonging to different patrilines would of course be half sisters, with a coefficient of genetic relatedness of 0.25. The fitness gained by rearing half sisters would obviously be considerably less than that gained by rearing offspring. In species such as the honey bee, where workers do not have the option of either leaving or of driving away the queen and taking over her role, the habit of multiple mating may, as we just saw, set the workers against each other and help the queen. But in species where the workers can revolt, this habit of multiple mating by the queens should decrease the propensity of the queen's daughters to remain in her nest and help her raise more daughters. The question therefore is: Why should queens mate with more than one male? Would they not be better off mating with a single male and thus ensuring the cooperation of their daughters? In search of a solution to this apparent paradox researchers have begun to focus on the possible advantages of genetic variability (provided by the presence of multiple patrilines) within a colony. For instance, to the extent that task performance in the colony has a genetic basis (and we saw evidence of this in Chapter 3), genetic variability provides for a more efficient division of labor. A somewhat different kind of argument is that intra-colony genetic variability could provide effective resistance to diseases, which might otherwise spread rapidly when all workers in a colony are highly related to each other and thus susceptible to the same parasites.

Jacqui Shykoff and Paul Schmid-Hempel studied the European bumble bee *Bombus terrestris* and its intestinal trypanosome parasite, *Crithidia bombi,* and confirmed such an advantage of intra-colony genetic variability. *C. bombi* spreads from one bumble bee to another through the ingestion by bumble bees of live parasite cells during direct physical contact or through contact with the feces of infected individuals. *B. terrestris,* living in a temperate environment, suspends

Drawing of a bumble bee nest showing brood in different stages of development, wax pots filled with honey, and adults. *(Drawing by Margrit Pirker.)*

colonial life during the winter, when new queens hibernate while old queens and all workers die. In the following spring, queens emerge from their hibernation and initiate new colonies. The queens first produce daughters who become workers and later, with the help of this labor force, they produce daughters who mate, hibernate, and become queens in the next year. The parasite depends for its continued survival through the years on infecting new queens before they begin to hibernate. Infected queens are likely to pass the infection on to their daughter workers as well as their daughter queens because of the possibility of becoming infected through physical contact or contact with feces within their colonies. Laboratory experiments show that the spread of infection from one bumble bee to another depends significantly on the genetic relatedness between the source and the recipient of infection. This suggests a genetic basis for susceptibility and supports the idea that infection would spread more rapidly in a relatively genetically homogeneous colony than in a colony of genetically more variable individuals. Given a reasonable chance of their being infected, queens who mate multiply and produce genetically variable daughters should therefore be at an advantage compared with queens who mate singly and produce genetically similar daughters. Disease is thus a potential factor that selects for multiple mating by the queen, although *B. terrestris* queens seem to mate singly, perhaps for the reason mentioned below.

The *Bombus-Crithidia* story has other fascinating ramifications. In normal uninfected colonies, workers at first have poorly developed ovaries and spend all their time working for the colony to rear the queen's (their mother's) brood. Over time, however, the workers gradually develop their ovaries, and toward the end of the colony cycle they virtually revolt against queen control and begin to lay their own eggs. The success of queens therefore depends upon producing new daughter queens before workers begin to revolt. A queen that dies after producing only workers and no daughter queens gains little, if any, fitness. Curiously, queens seem to benefit from the infected status of their daughter workers. Infection retards the ovarian development of workers and thus keeps them working for longer periods of time and postpones the time of their revolt. In these circumstances queens have more time to complete the production of their new daughter queens. In principle this should provide an opposing selective force. Since queens benefit from having infected workers and such infection spreads more effectively in genetically similar lines of workers, a queen would be better off mating singly and producing daughters who are all full sisters. Disease could thus in principle select for single mating, instead of multiple mating, and this may perhaps be the reason why the queens seem to mate singly.

But there is a problem here. If workers in a colony are infected, the new daughter queen is likely to be infected too. A parasite that has detrimental effects on workers has similar effects on the queens. Infected queens can start new nests, but they lay eggs at a somewhat lower rate than do uninfected queens. This has been shown to lead to significantly smaller worker populations in infected laboratory colonies. So the queens should prefer to avoid infection in their colonies and should mate with many males. Obviously we do not know which factor is more important and we therefore cannot say confidently why the queens mate singly. My only purpose here is to point out the various ways in which disease can influence phenomena like multiple mating, which in turn influence levels of intra-colony genetic relatedness. All this reasoning is from the queen's point of view, because selection for multiple versus single mating is expected to act on the queen—the daughter workers have little say in this matter.

But multiple mating is only one way of increasing genetic variability in the colony. The presence of multiple queens is another way. Here it is entirely possible that workers have some say in the matter. In some ants for example, it is well known that workers decide not only how many queens may be reproductively active in their colony but even which individuals may become reproductively active queens. Now what will the workers prefer—low genetic variability or high genetic variability? Disease can have profound and unexpected consequences for the balance between cooperation and conflict, but it remains a poorly studied aspect of social life in animals.

Chimpanzee Politics

In the 1960s, Jane Goodall went to Africa to undertake her path-breaking study of chimpanzees in the wild. In the 1970s, Allen and Beatrice Gardner worked with chimpanzees in captivity and taught them American sign language. In the 1970s and 1980s, Frans de Waal spent many years watching chimpanzees in a large outdoor enclosure in the Arnheim Zoo in the Netherlands. In many ways, de Waal's research, though conducted on animals in captivity, gives us a superb picture of chimpanzee behavior, because de Waal could avoid the difficulties of observation in the wild and yet continually watch chimpanzees whose enclosure was large enough to permit them to behave naturally. De Waal's most telling observations about these animals, recorded in *Chimpanzee Politics,* concern the love-hate relationships between three males, Yeroen, Luit, and Nickie. In the beginning, Yeroen was the dominant, or alpha, male. Luit and Nickie as well as all the females treated him with respect. Luit gradually challenged Yeroen by enlisting the cooperation of the females as well as of young Nickie. The very fact that Nickie was used by Luit to wrest power from Yeroen appears to have given Nickie an advantage. It was not long before Nickie, with help from none other than the ousted Yeroen, challenged and replaced Luit as the new alpha male. Not only did de Waal witness the swinging back and forth between cooperation and conflict between Yeroen, Luit, and Nickie, but on almost every day he witnessed conflict and reconciliation among the members of the group.

In his foreword to *Chimpanzee Politics,* Desmond Morris writes of the chimps: "Their life is full of takeovers, dominance networks, power struggles, alliances, divide-and-rule strategies, coalitions, arbitration, collective leadership, privileges and bargaining. There is hardly anything that occurs in the corridors of power of the human world that cannot

The three famous chimps Yeroen, Luit, and Nikkie (left to right). *(Reprinted with permission from F. de Waal, Chimpanzee Politics: Power and Sex among Apes, 1989).*

be found in embryo in the social life of a chimpanzee colony." Not surprisingly, de Waal contends that "the roots of politics are older than humanity." Morris suggests that this message "will upset many including some of our leading political figures." If that is so, I shudder to think of what my next section will do.

Wasp Politics?

In April 1981 I was studying a colony of *Ropalidia cyathiformis*. The colony began to show a steep decline in both the number of adults and the number of brood being reared. I feared that, as often happens, the colony might be abandoned, bringing a premature end to my long-term study. Instead, what actually happened was far more interesting. On the evening of May 31, 1982, I had left the colony with 11 adult females, all individually marked with spots of different-colored paint. On my arrival on the morning of June 1, I noticed with dismay that only 6 of the 11 females remained on the nest. It is not unusual for 1 or 2 wasps at a time to disappear from such colonies. But the disappearance of 5 wasps (nearly half the population) overnight aroused my suspicion. More than anything else, I did not want this colony to be abandoned. I really wanted to find the missing wasps. That did not take long. I had only to look around for a few minutes when, to my amazement, I found all 5 of the missing wasps, which I could identify with certainty by their paint spots. What amazed me more was that the 5 wasps were not just sitting there; they had a small nest of their own.

It then dawned upon me that these 5 wasps had deserted their original colony, perhaps revolting against the authority of the queen, and had decided to start their own new nest. It did not take me long to find out that Orange, one of the particularly aggressive individuals on the original nest, had become the queen on the new nest. My disappointment at the loss of half my wasps turned into great excitement. Clearly, half the population had deserted their declining colony and ventured out on their own. Perhaps the aggressive Orange had led the revolt and walked away with her followers. This event raised several questions in my mind. I could easily imagine that, being dissatisfied with the state of the original colony, but not being able to dislodge the original queen and mend matters, Orange decided to leave.

But what would be the consequence of this for the Rebels that left and indeed for the Loyalists that remained in the original colony? This was easy to determine. I simply continued my observations and included the new colony in my study. The result was remarkable. The colony fission turned out to be good for both the Rebels and the Loyalists. The Rebels did very well; their colony grew rapidly and they began to rear brood quite successfully. Even more remarkable, the Loyalists in the original colony also benefited. In sharp contrast to the declining condition of the colony before the fission, the situation there improved and they too began to rear brood quite successfully. Clearly, the fission increased the fitness of both the Rebels and the Loyalists. But why was there such a difference in the level of cooperation before and after fission? It was my impression that there was too much aggression on the nest before fission. A quantitative analysis of the behavior of the wasps before and after fission confirmed this suspicion.

An analysis of the pattern of aggression before the fission was even more instructive. Having witnessed the fission and identified the Loyalists and the Rebels, I could now go back to the behavioral data on these individuals in my computer files and compare the behavior of the Loyalists and the Rebels before the fission occurred. It turned out that the Loyalists were the real aggressors; they showed much more aggression toward the Rebels than the Rebels did toward them. Indeed, the Loyalists also appeared to have driven away a number of other individuals during April and May 1982, although I have no idea of the fate of these other individuals. It is reasonable to conclude therefore that high rates of aggression reflect a high degree of conflict and that this reduced the efficiency of brood rearing before colony fission. Conversely, the low rates of aggression in both colonies after fission reflect a high degree of cooperation and this allowed efficient brood rearing.

But how did the Rebels manage to get together and leave at the same time and reach the same site to start a new nest? Was it a snap decision taken on the night of May 31 or had revolt been brewing for some time? Was there some form of groupism even before the fission? To investigate these questions my colleagues and I measured behavioral coordination within and between subgroups (Rebels and Loyalists) using a mathematical index called Yule's association coefficient. We then asked whether there was more coordination within subgroups than between subgroups. For instance, did wasps within a subgroup synchronize their trips away from the nest and did Rebels and Loyalists avoid each other? It turned out that the Rebels had high association coefficients among themselves. Similarly, the Loyalists among themselves also had a positive association coefficient, although this was not as high as the value among the Rebels. In contrast, Rebels and Loyalists had a negative association with each other. This suggests that the wasps had differentiated into two subgroups well before the fission, with the Loyalists and Rebels behaving as two coordinated subgroups and avoiding each other. The wasps must therefore have been capable of individual recognition and must have had some way of deciding when to leave and where to go.

Do Wasps Form Alliances?

In early 1985 I had another nest under observation for the purpose of removing the queen to see who would be the next queen; indeed my long-term goal was to predict the identity of successors to ousted queens. The behavior of two of the wasps was particularly interesting. Red was very aggressive, and particularly so toward Blue. She would harass Blue so often and for such prolonged periods of time that on several occasions I noticed that the queen would intervene. The queen would actually climb on the grappling mass of Red and Blue and separate them. This was clearly of great help to Blue, who was no match for Red. I got the distinct impression that Blue was not only trying to avoid Red but also trying to appease the queen.

The most dramatic example of this occurred one day when Blue returned to the colony with food but before she could land on the nest, Red noticed her and poised herself to grab the food from Blue. It appeared that Blue did not want to give the food to Red. It also appeared that she wanted to give the food to the queen. But the queen was looking the other way and did not notice Blue arrive. Blue's response was very interesting. She landed on the leaf on which the nest was built about 2 centimeters away from the nest, something that returning foragers seldom do—they usually alight on the nest. Having done that, Blue sat on the leaf, and Red sat on the nest, and they went through what might be called a war of attrition for over 5 minutes; Blue made several attempts to get on the nest but Red always blocked her way and tried to grab the food. Having failed to attract the attention of the queen or to climb onto the nest without losing the food load to Red, Blue now simply walked around the nest and came in full view of the queen. The queen seemed to immediately sense what was going on. She let Blue climb onto the nest and took the food load from her mouth, but at the same time Red pounced on Blue and bit her. Before too long, Blue managed to escape from the clutches of Red and fly away.

This episode, dramatic as it already was, assumed even greater significance in light of what happened after I removed the queen. Clearly, Red was the next most dominant individual and I had little doubt she would be the next queen after I removed the present one. But to my surprise, it was Blue who became the next queen, in spite of Red's presence. Indeed, Red stayed in the colony for over a month after Blue took over, but I cannot help describing her behavior as "sulking"—she would do nothing at all except occasionally take some food from one of the foragers. She did not participate in any nest activity.

Why was Red so much more aggressive toward Blue than toward other individuals? Why was the queen so "considerate"of Blue? Was there some kind of alliance between Blue and the queen? If so, did it have any influence on Blue's becoming the next queen when I removed the original queen, even though Red was higher in the dominance hierarchy?

Do Wasp Workers Choose Their Queens?

During a similar queen-removal experiment with *Ropalidia cyathiformis,* I once had a situation when there were two contenders, as it were, to replace the existing queen. These were Blue and Orange (different from the Blue and Orange of the two previous stories), both more or less equally dominant. When I removed the queen on March 9, 1985, Blue took over the place of the queen and Orange promptly left the colony. Blue, however, was apparently not a very "good" queen. All the other wasps stopped foraging and began to simply sit on the nest. Even when they did go out, they returned with nothing. Clearly Blue had eggs to lay because she began to cannibalize on existing eggs to make room for her to lay her own, since no wasp would supply building material or build new cells for her. Eventually, other wasps began cannibalizing on brood too and I was afraid that the colony would be abandoned. I was amazed to notice, however, that Orange had not quite given up. She would occasionally come back to the nest, as if to check on how Blue was doing. She would never spend the night on the nest but would only visit occasionally. By about the March 20, Orange returned for good and Blue left. A pity that I was not there to witness their meeting! Now the behavior of the rest of the wasps was dramatically altered. They began to work—they foraged, brought food, fed larvae, extended the walls of the cells of the growing larvae, and even brought building material and built new cells for their new queen, Orange, to lay eggs in.

The story does not quite end there. Blue also, it turned out, had not quite left the nest. She would also come from time to time and visit, as if to see how her rival, Orange, was doing. After a few days Blue decided to rejoin the nest, but not before experiencing a great deal of hostility from the resident wasps. Blue had to spend nearly a whole day being subordinated by several residents before she was accepted back into the colony. Once again, we see that wasps can recognize individuals, and it also appears that they can modify their behavior based on that recognition. Why did the wasps not cooperate with Blue when she first took over as the queen? If she was simply not good enough to be a queen, why did she succeed in the first place, especially in the presence of Orange? Wasp politics?

Paternal Harassment of Sons in the White-Fronted Bee-Eater

Let us return to the study of the white-fronted bee-eater discussed earlier in Chapter 6. Emlen and Wrege saw white-fronted bee eaters engage in a bizarre kind of conflict. Some individuals, particularly adult males, harassed other members of their clan, particularly their sons, and prevented them from starting their own families. Harassment included persistently chasing potential breeders away from their territories, interfering with their courtship by preventing them from feeding their consorts, and physically preventing potential breeders from entering their nests by blocking the nest entrances. A frequent consequence of such behavior was that the harassed individual abandoned its attempts to breed and returned to the harasser's nest to act as a helper. Why do adult males harass potential breeders in this fashion? Why do they seem to particularly choose their sons as targets of harassment? Why do the sons accede to such harassment and not resist it more firmly? Why is it that the adult males have the greatest success in recruiting helpers through harassment when they target their sons?

Amazing as it may seem, all these apparent paradoxes are understandable within the framework of inclusive fitness theory. Since Emlen and Wrege had all their bee-eaters marked and the fate of each nest recorded, they could compute the costs and benefits of harassing as well as of acceding to harassment. First let us look at harassment from the point of view of the adult males. What are the costs and benefits of harassing their sons? If harassment is successful, the sons will come back to the nest as helpers and increase the number of offspring that the adult males can produce. That is a benefit. But then the sons will not breed on their own and hence the harasser will lose some grandchildren. On the average, a nest without helpers—with the only adults being the breeding pair—produced 0.51 offspring, while a nest with one additional helper produced 0.98 offspring. Fathers who harass their sons and bring them back would gain $0.98 - 0.51 = 0.47$ offspring and lose 0.51 grandchildren. Since 0.47 offspring are far more valuable than 0.51 grandchildren (remember the father is related to his offspring by 0.50 and to his grandchildren by only 0.25), natural selection should favor fathers who harass their sons.

But why does the son not resist? Let us now do the calculation from his point of view. A son who came back and helped his father would contribute to the production of 0.47 siblings and lose about the same number, 0.51, of offspring (that he might have produced on his own). Since he is equally related to his siblings and to his offspring (note that we are now dealing with a diploid system and a not a halpodiploid system, as occurs in the Hymenoptera), it does not matter too much to the son whether he helps or breeds. Thus natural selection on the son will not be very strong. The fathers will be selected to keep trying to get back their sons while sons will not be selected to resist too strongly.

Breeding Options in the White-Fronted Bee-Eater

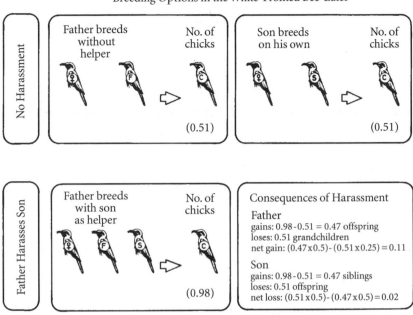

Breeding options in the white-fronted bee-eater. *(Based on Emlen and Wrege 1992; drawing by Sanjeeva Nayaka.)*

Communal Nursing in House Mice

House mice live in social units that typically consist of a single dominant male and one or more adult females with their offspring. The male mates with each female in his unit but provides no parental care to the pups. When there is more than one female in a reproductive unit, the females have abundant opportunities for cooperation and conflict, especially since they all litter at about the same time and rear their pups in a communal nest.

Barbara König at the University of Würzburg in Germany created monogynous and polygynous reproductive units in the laboratory to examine the balance between cooperation and conflict. On average, females in polygynous units fared better than those in monogynous units, especially if the females sharing a communal nest were sisters (sisterhood appears to be inferred by the mice on the basis of familiarity—a reasonable basis for such inference in nature, where sisters are likely to be familiar and nonsisters are likely to be unfamiliar). The main reason for the better performance of mothers rearing their pups in communal nests is that a given female suckles not only her own offspring but also those of her breeding partner—a behavior that human mothers would consider an extreme example of altruism. Perhaps even more striking to the biologist is the apparent inability of the females in a communal nest to discriminate between their own offspring and those of their partners, even when there is considerable age difference be-

Communal nursing in house mice: mothers nurse their own pups along with other pups (of their breeding partners), which may have reached a significantly more advanced stage of development than their own pups. *(Drawing: S. Bonsani.)*

tween their own and alien pups. Such communal nursing is observed even if the females living in a communal nest are unrelated and unfamiliar; clearly this is not merely another case of nepotism.

Females who nurse nonoffspring must gain something, and here is where the conflict comes in. The expression of conflict can be as dramatic as the expression of cooperation. Females who nest together do not litter at exactly the same time; their litters are usually separated by about 8 days. The female who litters later often kills one or two offspring of the female who litters first—the infanticide being committed here by females, whereas among the langurs it was always the males that were infanticidal. The probability of committing infanticide is higher if the female who has the opportunity to do so has a large litter inside her body—it appears that the mice have a way of assessing the litter size even before giving birth. Having killed one more of her partner's offspring, the female then litters and begins to nurse her own offspring and the remaining offspring of her partner indiscriminately, as if nothing unusual had happened. Production of milk is costly, and by killing one or two of her partner's offspring the infanticidal female causes increased flow of milk from the other female to her own offspring; obviously this is more important when her own litter is a large one.

Why does the second female not kill all the offspring of the first female and attempt to channel all her partner's milk to her own offspring? It appears that if the first female loses all her offspring she will cease to produce milk, and it would therefore not be prudent on the part of the second female to kill all the first female's offspring. Why don't females nest alone and avoid having their offspring killed in this fashion? Barbara König's experiments show convincingly that monogynous females produce fewer offspring than each member of the polygynous pair. The female who litters first and perhaps loses some of her offspring through infanticide by her partner will probably be the second to litter next time around, and will then be able to pay her partner back in the same coin. When you compute their lifetime reproductive success, the two females more or less break even with each other but both do better than monogynous females. By killing some and suckling some of their partner's offspring, by a mix of cooperation and conflict, these females achieve higher fitness than do solitary mothers. This example illustrates rather strikingly that you can rarely have cooperation without some conflict or conflict without some cooperation. Both cooperation and conflict are inevitable consequences of social life, and they are often inseparable components of a survival strategy.

When Ant Queens Mutilate Workers

Diacamma is a rather unusual ant. It lives in societies with a single reproductive that we might call the queen, with the remaining ants acting as workers with a fine division of labor. Since ants usually have a morphologically well differentiated queen, we can tell from observation that such a true queen does not exist in *Diacamma*. It appears that the workers have simply stopped producing queens and have learned to manage on their own. This is a group of ants where the workers have not lost the ability to mate and store sperm. So the workers can indeed manage on their own. In a way, then, these ant colonies are like my *Ropalidia* colonies, where one of the many nearly identical individuals assumes the role of a queen.

Christian Peeters and his colleagues carefully examined such queens (whom they called gamergates, to distinguish them from the morphologically differentiated true queens of other ants) and workers from species of *Diacamma* from southeast Asia and Australia, and found that the gamergates and workers are morphologically different after all. The gamergates have little wing buds called gemmae (ant workers do not

A typical nest mound of the queenless ant *Diacamma ceylonense* in the Indian Institute of Science, Bangalore. *(Photo: K. Kunte.)*

have wings), while the workers don't have gemmae. But what makes *Diacamma* remarkable is what happens after the ants are born—a colloquial expression for emerging from the pupal case; an insect should perhaps be said to be born twice, once when the egg is laid and again when it completes metamorphosis and emerges as a completely transformed individual from its pupal case. All individuals are born with their gemmae intact, but the gemmae of all the ants are physically and violently mutilated by the queen. If the queen dies, the first ant to emerge subsequently retains her gemmae because there is no one to

Scanning electron micrograph of *Diacamma australe* showing the thoracic appendages called gemmae. *(Reproduced with the permission of Springer-Verlag from C. Peeters and S. Higashi, "Reproductive Dominance Controlled by Mutilation in the Queenless Ant Diacamma australe," Naturwissenschaften, 76 [1989]: 178.)*

remove them, and she then systematically mutilates the gemmae of all who emerge after her. The gemmae are required for mating, probably because they send chemical signals to the males. Ants without gemmae are characteristically mild and submissive and workerlike, while ants with gemmae are dominant and aggressive and characteristically queen-like.

Nobody knows how the presence or absence of gemmae affects the behavior of the ants, but here is a system where queens maintain their status as the sole reproductive of the colony and suppress all other individuals by mutilating them. The system is so designed that if the gamergate dies accidentally, the next individual to emerge will automatically become the next gamergate. The workers who have had their gemmae mutilated appear to work efficiently for their colonies without any trace of discord, and may actually be selected by natural selection to do so owing to the indirect social component of inclusive fitness that they get by rearing the gamergate's offspring. Nevertheless, it is evident that conflict is the flip side of cooperation is evident from the fact that the "queens" have to mutilate workers to get them to work.

When Ant Workers Kill Queens

Solenopsis is a small ant with a very painful sting. If you accidentally step on one of its nest openings, you will soon have hundreds of ants crawling over you and biting you. You will soon feel that your body is on fire, and you will understand why these creatures are called fire ants. One species, *Solenopsis invicta,* occurs naturally in Argentina and has been accidentally introduced into North America. As a recent introduction, it has few or no enemies in the new habitat and is rapidly spreading across the southern United States. North Americans seem to consider the fire ant a serious pest and are pouring huge amounts of money into fire ant research. I have no complaints because this has resulted in some of the finest research into ant biology.

Kenneth Ross at the University of Georgia and a visiting scientist from Switzerland, Laurent Keller, have recently uncovered a fascinating aspect of the fire ant story. Some colonies have a single queen while others have many queens. Being monogynous or polygynous appears to be a matter of tradition (if wasps can indulge in politics, why can't ants have tradition?); monogynous colonies rear big fat queens, suitable for starting new monogynous colonies, while polygynous colonies rear small queens, suitable only for entering and surviving in other polygynous colonies. In polygynous colonies, workers seem to limit the food given to maturing queens and if they encounter a really strong queen (some queens appear to be capable of becoming strong and dominant by virtue of their genetic make-up), they kill her and thereby ensure that a single dominant queen does not bully all other queens into submission and convert the colony into a virtually monogynous one. Thus polygynous colonies cannot turn monogynous because their polygynous state is perpetuated by the workers, who will not let a single queen dominate. For this reason Ross and Keller have described polygyny in *Solenopsis* as a "culturally" transmitted character, one passed on from one generation to another irrespective of the genetic make-up of the queens that enter an already polygynous colony.

Who's the Boss?

In most highly developed ant and honey bee colonies, the queen normally produces one or more chemical substances, called pheromones, that are meant to suppress the workers and prevent them from developing their ovaries and laying eggs. This quite naturally suggests that the queen controls the workers for selfish reasons and that the workers are forced to behave in an apparently altruistic manner. We then go on to explain that the workers' altruism is not eliminated by natural selection if they gain sufficient inclusive fitness by rearing the queen's offspring, who may be their relatives. But this also means that the worker is acting selfishly by preferring to be a sterile worker rather than going off on her own to start a new nest because staying gives her more inclusive fitness than leaving. So who's the boss in the ant or bee colony? Is the queen controlling the workers or are they staying "voluntarily"? This is not just a matter of semantics. We cannot define the function of the queen pheromones until we decide who's the boss. All along, we have thought of the queen as the boss and regarded the queen pheromones as worker-controlling substances.

Laurent Keller and Peter Nonacs have recently challenged this view and argued that we must think of the queen pheromones not as substances meant to control the workers but as signals used by the workers to voluntarily curtail their own reproduction because they are better off if the queen reproduces. One interpretation is that the workers are prevented by the queen from reproducing and that in her absence this inhibition is removed and they start reproducing. A different interpretation is that the workers prefer having the queen reproduce rather than doing so themselves because the queen is so much better at it. But if the queen dies, it is better for them to reproduce than for nobody to do so. Hence they use the queen pheromones as a signal to decide whether they should let the queen continue to lay eggs or whether they should do so themselves. So who's the boss? Perhaps the question is a pointless one, after all. From the point of view of natural selection, there is no boss; each individual is attempting to maximize its own inclusive fitness and the net result is that there is always a precarious balance between cooperation and conflict.

Little wonder then that the fine balance between cooperation and conflict is ubiquitous. Although slime mold amoebae are willing to commit suicide to enable some of them to disperse, they are always ready to cheat if some of the members of the group are genetically unrelated. Both parents of the blue tit family are willing to feed the chicks in apparent harmony, but the male is always ready to quit after bringing up a few strong chicks and the female has her own way of ensuring his continued cooperation by making all chicks the same age and size. Worker ants and bees spend their whole life caring for the queen's offspring but will try to feed their sisters more food than their brothers. Worker honey bees will try to sneak in their own sons in place of brothers, but the queen retaliates, creating discord among the workers by ensuring that they are only half sisters. Queen bumble bees will even make their daughter workers more susceptible to disease if that's what it takes to ensure their prolonged cooperation. Queens will mutilate workers and workers will kill queens if that's what it takes to ensure "harmonious" social life.

I have endeavored to show that both cooperation and conflict are inevitable consequences of the survival strategies of social animals and that a common theoretical framework can be developed to explain the observed mix of cooperation and conflict in different situations, whether we are dealing with slime molds or with chimpanzees. The examples I have chosen are my favorites, but cooperation and conflict are endemic in all animal societies.

10

Some Caveats and Conclusions

The Power of Simplifying Assumptions

A point that I have emphasized right from the beginning is that physiologists and evolutionary biologists should not quarrel about whose explanation is more correct. If the physiologist finds that birds migrate because their pineal gland has detected changes in day length and the evolutionary biologist finds that the cost of migration is less than the cost of having to spend the winter in the northern latitudes, both are correct because they are dealing with two different levels of explanation. It does not make sense to try to decide which of the two explanations is better. Much unnecessary debate and confusion is avoided if we recognize the distinctness of the two different levels of analysis and work within either one of them. Ignoring the possible physiological explanations and focusing on the evolutionary explanation or vice versa appears to be a legitimate way of avoiding confusion. It is also sometimes inevitable because the training and methodology and quite often the very philosophy of scientific research underlying the physiological and evolutionary explanations may be quite distinct. But the time must come in

171

the development of this field of scientific inquiry when we begin to integrate the physiological and evolutionary explanations. After all, we agree that both explanations are correct at their respective levels. A little reflection should convince us that such integration will eventually become essential for the further pursuit of each of the two levels of inquiry.

Let us take a specific example. We saw in Chapter 7 that all wasps are not born equal and that this makes it easier for some of them to adopt the worker strategy while others adopt the solitary nesting strategy. We also saw that variation in larval nutrition was the basis of such differentiation between the egg layers and non–egg layers. Let us for the sake of convenience call the egg layers potential queens and the non–egg layers potential workers. We recognized that differential larval nutrition could result from two different mechanisms: (1) accidental variations in the amount of food given to the larvae on account of the inability of the adult wasps to apportion food accurately and (2) an evolved ability of the workers to "deliberately" feed some larvae more and others less. We labeled these proximate mechanisms and decided to ignore them because as far as the evolutionary consequences are concerned, it does not matter which is operative. Differential larval nutrition can facilitate the evolution of a worker strategy irrespective of whether it results from accidental variation in the amounts of food given to larvae or whether it occurs because workers have the ability to apportion food unequally between groups of larvae. We did not need to wait for the discovery of the exact mechanism of generating differential larval nutrition before going on to explore the evolutionary consequences (and even if we did understand the proximate mechanism we would want to know about its broader consequences). It therefore seemed reasonable to ignore the proximate mechanism and focus on the evolutionary consequences. This does not mean that understanding the proximate mechanism is unnecessary in any general sense. Indeed, understanding one level of explanation can often be very helpful in further exploring the other level. In this case understanding the proximate mechanism by which larvae are differentially fed will considerably help in exploring the possible evolution of the worker strategy through differential larval nourishment.

For example, if larvae are differentially nourished because the workers regulate the amount of food given to different larvae, the adult wasps at least potentially have the ability to produce queens and workers at will, or, at the very least, they have the ability to skew the ratio of queens to workers in any desired direction. The ability to skew queen-worker ratios in response to environmental factors may significantly speed up the evolution of sociality. By contrast, if differential larval nourishment is the result of accidental fluctuations in larval feeding rates, the evolution of workers through this mechanism will probably be a slow and relatively inefficient process. However, there will then be a different kind of consequence. If *Ropalidia marginata* workers have the ability to deliberately feed different larvae different quantities of food, then it may well be that workers evolve first by some other mechanism and then later develop the ability to feed larvae differentially and thereby speed up the process of the evolution of sociality. If, on the other hand, *R. marginata* workers do not have the ability to feed workers differentially and it is accidental fluctuations in food given to different larvae that determine the differentiation into potential workers and potential queens, then we can conclude that accidental variation in larval feeding rates can potentially give rise to queen-worker differentiation so long as other conditions are appropriately conducive. A knowledge of the proximate mechanism of differential larval nourishment can thus greatly enhance the sophistication of our evolutionary explorations. But it is equally important that in the beginning of our inquiry we do not let ignorance of the proximate mechanisms prevent us from exploring evolutionary explanations. Even more important, knowledge of the proximate mechanism should not preclude inquiry about evolutionary explanations. The simplifying assumption that the two levels of explanation are independent is useful in the beginning, but should eventually be discarded in favor of an integration of both.

Let us now recall several other such simplifying assumptions that we made in the course of our discussion and remind ourselves that these need to be relaxed at some point. Levels of natural selection is another example. When the fallacy of naive Wynne-Edwardian group selection became obvious, it was useful to assume that natural selection always acts at the level of the individual, and that was justified because it got people out of the habit of blindly invoking the good of the group. Today we realize that natural selection can, in principle, act at the level of the group—and indeed at any other level of biological organization—but this usually requires very special and unusual conditions. The assumption that natural selection does not usually operate at levels other than the individual has helped make arguments about selection at those levels appropriately sophisticated; these arguments now bear no resemblance to earlier arguments that invoked natural selection at any level that seemed convenient. But this preference for explanations at the individual level should not serve to impede progress in our understanding of genuine cases of natural selection at all levels of biological organization. That natural selection acts at the level of the individual is a simplifying assumption that needs to be cautiously relaxed by examining the merits of each case (see Chapter 2).

Kin recognition is yet another example of our strategy of making simplifying assumptions in the initial phases of study and relaxing them later. The complete absence of empirical evidence for kin recognition for 15 years after Hamilton proposed the theory of kin selection did not impede the acceptance of his ideas. Indeed, the field of sociobiology, based primarily on Hamilton's idea that natural selection should favor individuals who aid genetic relatives even if by doing so they sacrifice personal reproduction and risk their own lives was well developed by 1979, when Greenberg's experiments with sweat bees provided the first convincing evidence of kin recognition in animals (see Chapter 6).

Even to this day we make the simplifying assumption that animals are not conscious of their actions because we do not have good evidence that they are; even more important, conscious actions are not necessary for our theories of animal behavior to work. But this is one example where the simplifying assumption is, I am afraid, threatening to stifle the development of the study of animal consciousness. The complacency generated by the realization that we do not need animal consciousness to explain even the most sophisticated behaviors of animals, coupled with the real problems associated with defining and unambiguously demonstrating its existence, has made people far too skeptical of any suggestion that animal consciousness does exist. This is where there is urgent need to reiterate that simplifying assumptions are only a necessary evil and we should not allow them to block further progress. Of course it requires a bold man like Donald Griffin to swim against the current and lead the way (see Chapter 8), but whether that will be sufficient, only time will tell.

My final example of a simplifying assumption and perhaps the most important one concerns what we might call the gene-culture continuum. Animal behavior may be determined by one or a small number of genes, it may be the result of instinct, it may result from simple learning by each individual, or it may involve more complex cultural transmission of learned information from generation to generation. The simplifying assumption we now make is that as long as behavior has a genetic component which is sufficient to make it possible for natural selection to favor some forms of the behavior and disfavor other forms, it really doesn't matter how the behavior is determined. This simplifying assumption makes it possible for us to use the same language to speak about how natural selection favors the melanistic *carbonaria* over the peppered *typica* moths in a polluted environment (see Chapter 2); how natural selection favors slime mold strains that selfishly have a high spore to stalk ratio while in the company of unrelated strains (see Chapters 1 and 6); how natural selection favors blackcaps that migrate from Germany and Austria to England rather than to southern Spain in the winter (see Chapter 3); how natural selection favors wasps that remain on their mother's nest, work for a while, and then seize an opportune moment to displace her and take over as the next queen (see Chapter 7); how natural selection favors bee-eater fathers who harass their sons and bring them back to their own nests to work as nonreproductive helpers (see Chapter 9); and indeed how natural selection favors chimps like Nickie, who initially helped Luit displace Yeroen from the dominant position but soon sought the help of the ousted Yeroen to challenge and replace Luit and install himself as the new alpha male (see Chapter 9).

The purpose of the simplifying assumption is to permit us to explore the evolutionary consequences of different behavioral strategies. Needless to say, the manner in which genes are likely to influence these different kinds of behavior or other traits is expected to vary widely. This variation is not likely to be trivial or irrelevant to the operation of natural selection. Almost certainly the rate of natural selection will be greatly influenced by the extent to which genes influence behaviors. We are by no means blind to these complications when we talk in the same breath of slime mold selfishness and honey bee selfishness, when we talk of chimpanzee politics and wasp politics, or when we talk of culture in ants. A time will surely come when we will be in a position to explicitly study how the operation of natural selection is influenced by these different ways in which behavior is controlled and modulated, but we are not quite ready for that yet. But again, we should not let this lack of knowledge prevent us from exploring the evolutionary consequences of alternative behavioral strategies.

Where Do We Go From Here?

After a survey of our current state of appreciation of the evolution of cooperation and conflict in animal societies, it seems appropriate to ask, What next? One can speculate with reasonable confidence about some of the directions that inquiry into the evolutionary basis of cooperation and conflict is likely to take in the foreseeable future, although it would be foolish to try to forecast all possible developments. The simplifying assumptions that we have made, to serve as stepping stones for initial rapid progress, provide us with an obvious set of clues about the future. One can hope that these simplifying assumptions will be progressively relaxed. We have already seen this happening in the case of kin recognition and, to a small extent, in the case of levels of natural selection, and I hope this will also happen with other assumptions.

An area of study where one might reasonably expect major advances is the influence of genes on behavior. Continuing research on genetic predispositions for different behaviors in the honey bee and other animals and attempts to understand the mechanism by which genes influence behavior hold great promise. These expected advances will not only make sociobiological models of animal behavior more realistic but will probably open up completely new fields of research on how genes may influence behavior. It can be said with reasonable certainty that this enterprise will become increasingly multidisciplinary and bring together neurobiology, neural network models, ethology, and of course genetics, especially quantitative genetics.

Another area in which we might expect spectacular advances is the measurement of genetic relatedness. The power of modern molecular biological techniques is being increasingly used to determine the exact levels of genetic relatedness between interacting individuals. DNA fingerprinting and the use of hypervariable DNA sequence probes is permitting a level of sophistication in this aspect of the study of animal behavior that could not have been imagined just a few years ago. Thus, while the relatedness parameter in Hamilton's rule will soon be estimated to a high degree of accuracy, it is the estimates of the cost and benefit parameters that are going to become increasingly important—and worrisome. There is no new technology in sight that is likely to revolutionize the measurement of costs and benefits of animal behavior. We will therefore have to continue to depend largely on painstaking field and laboratory research using traditional methods. I worry that such traditional research will begin to lose out to the more attractive molecular approach, which unfortunately cannot give us all the answers by itself. That such molecular methods require large financial investments ironically appears to make them more and not less attractive to young researchers beginning their careers in evolutionary biology. The current short-sighted practice of preferring those who use molecular techniques for faculty appointments in evolutionary biology will unfortunately act as a powerful force of natural selection. In the years to come, our greatest challenge will be to maintain a healthy balance in investigating the different components that make up a theory of animal behavior.

Variety and diversity are the hallmarks of biological systems and with many millions of species we must be prepared to find that there are many different ways that animals have developed for achieving a given objective. This variation can easily be construed as a deterrent to developing unified theories, because there will always be more exceptions than conformations to any rule that we might attempt to discern. But biological variation also provides a unique opportunity to strengthen our theories because we can build into any theory predictions about how the expected pattern of behavior is likely to change with every changed circumstance. Biological diversity provides a plethora of nature's experiments that can all be used to test detailed predictions of our theories—the more exceptions to a general rule that we can discover and explain, the more mature will our theories get. Whatever the extent of variation, however, we can be certain that achieving a fine balance between cooperation and conflict is an invariant feature of the survival strategies of social animals.

Suggested Readings

Ali, S. 1985. *The Fall of a Sparrow*. Delhi: Oxford University Press.

Axelrod, R. 1984. *The Evolution of Co-operation*. New York: Penguin Books.

Dawkins, R. 1989. *The Selfish Gene*, 2nd ed. Oxford: Oxford University Press.

————1986. *The Blind Watchmaker*. London: Longman Group.

Diamond, J. 1991. *The Rise and Fall of the Third Chimpanzee*. London: Vintage Books.

Frisch, K. von. 1971. *Bees: Their Vision, Chemical Senses, and Language*. Ithaca: Cornell University Press.

Gould, J. L. 1982. *Ethology: The Mechanisms and Evolution of Behavior*. New York: Norton.

Griffin, D. R. 1992. *Animal Minds*. Chicago: University of Chicago Press.

Haldane, J. B. S. 1985. *On Being the Right Size and Other Assays*, ed. John Maynard Smith. Oxford: Oxford University Press.

Hölldobler, B., and E. O. Wilson. 1994. *Journey to the Ants*. Cambridge, Mass.: Harvard University Press.

Hrdy, S. B. 1977. *The Langurs of Abu: Female and Male Strategies of Reproduction*. Cambridge, Mass.: Harvard University Press.

Krebs, J. R., and N. B. Davies. 1991. *Behavioural Ecology: An Evolutionary Approach*, 3rd ed. Oxford: Blackwell Scientific Publications.

————1993. *An Introduction to Behavioural Ecology*, 3rd ed. Oxford: Blackwell Scientific Publications.

Maynard Smith, J. 1958. *The Theory of Evolution*. New York: Penguin Books.

Schaller, G. B. 1967. *The Deer and the Tiger*. Chicago: University of Chicago Press.

————1972. *The Serengeti Lion*. Chicago: University of Chicago Press.

Sunquist, F., and M. Sunquist. 1988. *Tiger Moon.* Chicago: University of Chicago Press.

Trivers, R. 1985. *Social Evolution.* Benjamin Cummings Publishing Co.

Williams, G. C. 1966. *Adaptation and Natural Selection: A Critique of Some Current Evolutionary Thought.* Princeton: Princeton University Press.

de Waal, F. 1989. *Chimpanzee Politics: Power and Sex among Apes.* Baltimore: The Johns Hopkins University Press. First published in 1982.

Wills, G. 1991. *The Wisdom of the Genes: New Pathways in Evolution.* Oxford: Oxford University Press.

Wilson, E. O. 1971. *The Insect Societies.* Cambridge, Mass.: The Belknap Press of Harvard University Press.

———1975. *Sociobiology: The New Synthesis.* Cambridge, Mass.: The Belknap Press of Harvard University Press.

———1978. *On Human Nature.* Cambridge, Mass.: Harvard University Press.

References

....

1. What Are Social Animals?

Ali, S. 1979. *The Book of Indian Birds,* 11th ed. Bombay: Bombay Natural History Society.

———1985. *The Fall of a Sparrow.* Delhi: Oxford University Press.

Ali, S., and S. D. Ripley. 1987. *Compact Handbook of the Birds of India and Pakistan,* 2nd ed. Delhi: Oxford University Press.

Bonner, J. T. 1967. *The Cellular Slime Molds,* 2nd ed., Princeton: Princeton University Press.

Clements, A. N. 1963. *The Physiology of Mosquitoes.* Oxford: Pergamon Press.

Frisch, K. von. 1971. *Bees: Their Vision, Chemical Senses, and Language.* Ithaca: Cornell University Press.

Gould, J. L. 1982. *Ethology: The Mechanisms and Evolution of Behavior.* New York: Norton.

Grinnell, J., C. Packer, and A. E. Pusey. 1995. Cooperation in male lions: Kinship, reciprocity, or mutualism? *Anim. Behav.,* 49: 95–105.

Gwinner, E. 1991. Internal rhythms in bird migration. In *Behavior and Evolution of Birds,* ed. Douglas W. Mock, pp. 20–31. New York: W. H.Freeman.

Olive, L. S. 1970. Mycetozoa: A revised classification. *Bot. Rev.,* 36: 59–87.

Schaller, G. B. 1967. *The Deer and the Tiger.* Chicago: University of Chicago Press.

———1972. *The Serengeti Lion.* Chicago: University of Chicago Press.

Sunquist, F., and M. Sunquist. 1988. *Tiger Moon.* Chicago: University of Chicago Press.

Winston, M. L. 1987. *The Biology of the Honey Bee.* Cambridge, Mass.: Harvard University Press.

2. Evolution, the Eternal Tinkerer

Brakefield, P. M. 1987. Industrial melanism: Do we have the answers? *Trends in Ecol. & Evol.*, 2: 117–122.

Cook, L. M., G. S. Mani, and M. E. Varley. 1986. Postindustrial melanism in the peppered moth. *Science,* 231: 611–613.

Darwin, C. [1859] 1964. *On the Origin of Species: A Facsimile of the First Edition.* Introduction by Ernst Mayr. Cambridge, Mass.: Harvard University Press.

Dawkins, R. 1989. *The Selfish Gene,* 2nd ed. Oxford: Oxford University Press.

———1986. *The Blind Watchmaker.* London: Longman Group.

Grant, B. S., D. F. Owen, and C. A. Clark. 1996. Parallel rise and fall of melanic peppered moths in America and Britain. *J. Hered.,* 87: 351–357.

Hrdy, S. B. 1977. *The Langurs of Abu: Female and Male Strategies of Reproduction.* Cambridge, Mass.: Harvard University Press.

Kettlewell, B. 1973. *The Evolution of Melanism.* Oxford: Clarendon Press.

Lorenz, K. 1963. *On Aggression.* New York: Bantam Books.

Mani, G. S. 1990. Theoretical models of melanism in *Biston betularia*—A review. *Biol. J. Linnean Soc.,* 39: 355–371.

Maynard Smith, J. 1964. Group selection and kin selection. *Nature,* 201: 1145–1147.

Mohnot, S. M., M. Gadgil, and S. C. Makwana. 1981. On the dynamics of the hanuman langur populations of Jodhpur (Rajasthan, India). *Primates,* 22: 182–191.

Nur, U., J. H. Werren, D. G. Eickbush, W. D. Burke, and T. H. Eickbush. 1988. A "selfish" B chromosome that enhances its transmission by eliminating the paternal genome. *Science,* 240: 512–514.

Perrins, C. 1964. Survival of young swifts in relation to brood-size. *Nature,* 201: 1147–1148.

Solbrig, O. T., and D. J. Solbrig. 1979. *Introduction to Population Biology and Evolution.* Reading, Mass.: Addison-Wesley.

Werren, J. H. 1991. The paternal sex-ratio chromosome of *Nasonia. Amer. Natur.,* 137: 392–402.

Werren, J. H. , S. W. Skinner, and A. M. Huger. 1986. Male-killing bacteria in a parasitic wasp. *Nature,* 231: 990–992.

Williams, G. C. 1966. *Adaptation and Natural Selection: A Critique of Some Current Evolutionary Thought.* Princeton: Princeton University Press.

————ed. 1971. *Group Selection.* Chicago: Aldine-Atherton.

Wynne-Edwards, V. C. 1962. *Animal Dispersion in Relation to Social Behaviour.* New York: Hafner Publishing Co.

————1964. Group selection and kin selection—a reply. *Nature,* 201: 1148–1149.

3. It's in the Genes

Ali, S. 1979. *The Book of Indian Birds.* 11th ed. Bombay: Bombay Natural History Society.

————1985. *The Fall of a Sparrow.* Delhi: Oxford University Press.

Ali, S., and S. D. Ripley. 1987. *Compact Handbook of the Birds of India and Pakistan,* 2nd ed. Delhi: Oxford University Press.

Berthold, R., A. J. Helbig, G. Mohr, and U. Querner. 1992. Rapid microevolution of migratory behaviour in a wild bird species. *Nature,* 360: 668–669.

Berthold, R., and U. Querner. 1981. Genetic basis of migratory behaviour in European warblers. *Science,* 212: 77–79.

Cade, W. H. 1975. Acoustically orienting parasitoids: Fly phonotaxis to cricket song. *Science,* 190: 1312–1313.

————1981. Alternative male strategies: Genetic differences in crickets. *Science,* 212: 563–564.

Gwinner, E. 1991. Internal rhythms in Bird migration. In *Behavior and Evolution of Birds,* ed. Douglas W. Mock, pp. 20–31. New York: W. H. Freeman.

Hall, J. C. 1985. Genetic analysis of behavior in insects. In *Comprehensive Insect Physiology, Biochemistry, and Pharmacology,* ed. G. A. Kerkut and L. I. Gilbert. New York: Pergamon Press.

————1990. Genetics of circadian rhythms. *Ann.Rev. Genet.,* 24: 659–697.

Hardin, P. E., J. C. Hall, and M. Rosbash. 1992. Circadian oscillations in period gene mRNA levels are transcriptionally regulated. *Proc. Natl. Acad. Sci. USA,* 89: 11711–11715.

Konopka, R. J. and S. Benzer 1971. Clock mutants of *Drosophila melanogaster. Proc. Natl. Acad. Sci.USA,* 68: 2112–2116.

Krebs, J. R., and N. B. Davies. 1993. *An Introduction to Behavioural Ecology,* 3rd ed. Oxford: Blackwell Scientific Publications.

Page, R. E. Jr. and G. E. Robinson 1991. The Genetics of Division of Labour in Honey bee Colonies. *Advances in Insect Physiology,* 23: 118–169.

Page, T. L. 1994. Time is the Essence: Molecular analysis of the biological clock. *Science,* 263: 1570–1572.

Robinson, G. E. 1992. Regulation of division of labor in insect societies. *Ann. Rev. Entomol.,* 37: 637–665.

Robinson, G. E., and R. E. Page, Jr. 1988. Genetic determination of guarding and undertaking in honey-bee colonies. *Nature,* 333: 356–358.

Rothenbuhler, W. C. 1964. Behavior genetics of nest cleaning in honey bees. *Amer. Zool.,* 4: 111–123.

Sehgal, A., J. L. Price, B. Man, and M. W. Young. 1994. Loss of circadian behavioral rhythms and per RNA oscillations in the *Drosophila* mutant timeless. *Science,* 263: 1603–1606.

Subbaraj, R., and M. K. Chandrashekaran. 1977. "Rigid" internal timing in the circadian rhythm of flight activity in a tropical bat. *Oecologia,* 29: 341–348.

———1978. Pulses of darkness shift the phase of a circadian rhythm in an insectivorous bat. *J. Comp. Physiol.,* 127: 239–243.

Sutherland, W. J. 1992. Genes map the migratory route. *Nature,* 360: 625–626.

Takahashi, J. S. 1992. Circadian clock genes are ticking. *Science,* 258: 238–240.

Wilson, E. O. 1975. *Sociobiology: The New Synthesis.* Cambridge, Mass.: The Belknap Press of Harvard University Press.

4. What Do Social Animals Do to Each Other?

FitzGerald, G. J. 1992. Egg cannibalism by sticklebacks: Spite or selfishness? *Behav. Ecol. Sociobiol.,* 30: 201–206.

Gadagkar, R. 1993. Can animals be spiteful? *Trends in Ecology and Evolution,* 8: 232–234.

James, C. D., M. T. Hoffman, D. C. Lighfoot, G. S. Forbes, and W. G. Whitford. 1994. Fruit abortion in *Yucca elata* and its implications for the mutualistic association with yucca moths. *Oikos,* 69: 207–216.

Linhart, Y. B., and R. J. Dodd. 1994. Yucca sex. *Nature,* 370: 604.

Moore, P. D. 1994. The yucca expediency. *Nature,* 368: 588–589.

Pellmyr, O., and C. J. Huth. 1994. Evolutionary stability of mutualism between yuccas and yucca moths. *Nature,* 372: 257–260.

Sherman, P. W. 1977. Nepotism and the evolution of alarm calls. *Science*, 197: 1246–1253.

Wilson, E. O. 1975. *Sociobiology: The New Synthesis.* Cambridge, Mass.: The Belknap Press of Harvard University Press.

5. The Paradox of Altruism

Darwin, C. [1859] 1964. *On the Origin of Species: A Facsimile of the First Edition,* Introduction by Ernst Mayr. Cambridge, Mass.: Harvard University Press.

Haldane, J. B. S. 1985. *On Being the Right Size and Other Essays,* ed. John Maynard Smith. Oxford: Oxford University Press.

Hamilton, W. D. 1964. The genetical evolution of social behaviour. *J. Theor. Biol.,* 7: 1–52.

Prete, F. R. 1990. The conundrum of the honey bees: One impediment to the publication of Darwin's theory. *J. Hist. Biol.,* 23: 271–290.

Richards, R. 1983. Why Darwin delayed, or interesting problems and models in the history of science. *J. Hist. Behav. Sci.,* 19: 271–290.

Sridhar, S., and K. P. Karanth. 1993. Helpers in the cooperatively breeding small green bee-eater *(Merops orientalis). Curr. Sci.,* 65: 489–490.

Trivers, R. L. 1971. The evolution of reciprocal altruism. *Quart. Rev. Biol.,* 46: 35–57.

Wilkinson, G. S. 1984. Reciprocal food sharing in the vampire bat. *Nature,* 308: 181–184.

———1986. Social grooming in the common vampire bat, *Desmodus rotundus. Anim. Behav.,* 34: 1880–1889.

———1988. Reciprocal altruism in bats and other mammals. *Ethol. and Sociobiol.,* 9: 85–100.

6. Do Animals Favor Their Relatives?

Badcock, C. 1990. What use is "euxeny"? *Nature,* 345: 391.

Bonner, J. T. 1967. *The Cellular Slime Molds,* 2nd ed., Princeton: Princeton University Press.

DeAngelo, M. J., V. M. Kish, and S. A. Kolmes. 1990. Altruism, selfishness, and heterocytosis in cellular slime molds. *Ethol. Ecol. Evol.,* 2: 439–443.

Emlen, S. T., and Wrege, P. H. 1989. A test of alternate hypotheses for helping behavior in white-fronted bee-eaters of Kenya. *Behav. Ecol. Sociobiol.*, 25: 303–319.

Gadagkar, R. 1985. Kin recognition in social insects and other animals: A review of recent findings and a consideration of their relevance for the theory of kin selection. *Proc. Indian Acad. Sci. (Anim. Sci.)*, 94: 587–621.

Gadagkar, R., and J. T. Bonner. 1994. Social insects and social amoebae. *J. Biosci.*, 19: 219–245.

Gadagkar, R., and A. B. Venkataraman. 1990. Nepotistic bee-eaters. *Curr. Sci.*, 59: 445–446.

Greenberg, L. 1979. Genetic component of bee odor in kin recognition. *Science*, 206: 1095–1097.

Hölldobler, B., and E. O. Wilson. 1977. Weaver ants. *Sci. Amer.*, 237(6): 146–154.

————1990. *The Ants*. Cambridge, Mass. The Belknap Press of Harvard University Press.

Maynard Smith, J., and M. G. Ridpath. 1972. Wife sharing in the Tasmanian native hen *Tribonyz mortierii*: A case of kin selection? *Amer. Natur.*, 106: 447–452.

Nanjundiah, V. 1985. The evolution of communication and social behaviour in *Dictyostelium discoideum. Proc. Indian Acad. Sci. (Anim. Sci.)*, 94: 639–653.

Packer, C., D. A. Gilbert, A. E. Pusey, and S. J. O'Brien. 1991. A molecular genetic analysis of kinship and cooperation in African lions. *Nature*, 351: 562–565.

Packer, C., D. Scheel, and A. E. Pusey. 1990. Why lions form groups: food is not enough. *Amer. Nat.*, 136: 1–19.

Rasmont, R. 1990. Immoral wasps? *Nature*, 344: 498.

Sherman, P. W. 1977. Nepotism and the evolution of alarm calls. *Science*, 197: 1246–1253.

Sridhar, S., and K. P. Karanth. 1993. Helpers in the cooperatively breeding small green bee-eater *(Merops orientalis). Curr. Sci.*, 65: 489–490.

Visscher, P. K. 1986. Kinship discrimination in queen rearing by honey bees *(Apis mellifera). Behav. Ecol. Sociobiol.*, 18: 453–460.

Wilson, E. O., and B. Hölldobler. 1980. Sex differences in cooperative silk-spinning by weaver ant larvae. *Proc. Natl. Acad. Sci. USA*, 77: 2343–2347.

7. A Primitive Wasp Society

Gadagkar, R. 1985. Evolution of insect sociality: A review of some attempts to test modern theories. *Proc. Indian Acad. Sci. (Anim. Sci.)*, 94: 309–324.

———1990. Evolution of eusociality: The advantage of assured fitness returns. *Phil. Trans. Roy. Soc. London, B.*, 329: 17–25.

———1990. Evolution of insect societies: Some insights from studying tropical wasps. In *Social Insects: An Indian Perspective*, ed. G. K. Veeresh, A. R. V. Kumar, and T. Shivashankar, pp. 129–152. Bangalore: IUSSI—Indian Chapter.

Gadagkar, R. 1991. *Belonogaster, Mischocyttarus, Parapolybia*, and independent founding *Ropalidia*. In *The Social Biology of Wasps*, ed. K. G. Ross and R. W. Matthews, pp. 149–190. Ithaca: Cornell University Press.

———1995. Cooperation and conflict in an insect society. *J. Indian Inst. Sci.*, 75: 333–352.

Gadagkar, R., S. Bhagavan, K. Chandrashekara, and C. Vinutha. 1991. The role of larval nutrition in pre-imaginal biasing of caste in the primitively eusocial wasp *Ropalidia marginata* (Hymenoptera: Vespidae). *Ecol. Entomol.*, 16: 435–440.

Gadagkar, R., K. Chandrashekara, S. Chandran, and S. Bhagavan, 1993. Serial polygyny in a primitively eusocial wasp: Implications for the evolution of sociality. In *Queen Number and Sociality in Insects*, ed. L. Keller, pp. 187–214. Oxford: Oxford University Press.

Gadagkar, R., C. Vinutha, A. Shanubhogue, and A. P. Gore. 1988. Pre-imaginal biasing of caste in a primitively eusocial insect. *Proc. Roy. Soc. London, B.*, 233: 175–189.

Muralidharan, K., M. S. Shaila, and R. Gadagkar. 1986. Evidence for multiple mating in the primitively eusocial wasp *Ropalidia marginata* (Lep.) (Hymenoptera: Vespidae). *J. Genet.*, 65: 153–158.

West-Eberhard, M. J. 1975. The evolution of social behavior by kin selection. *Quart. Rev. Biol.*, 50: 1–33.

Wilson, E. O. 1971. *The Insect Societies*. Cambridge, Mass.: The Belknap Press of Harvard University Press.

8. Games Animals Play

Axelrod, R. 1984. *The Evolution of Co-operation*. New York: Penguin Books.

Axelrod, R., and W. D. Hamilton. 1981. The evolution of cooperation. *Science,* 211: 1390–1396.

Davies, N. B. 1978. Territorial defence in the speckled wood butterfly *(Pararge aegeria):* The resident always wins. *Anim. Behav.,* 26: 138–147.

De Waal, F. 1989. *Chimpanzee Politics: Power and Sex among Apes.* Baltimore: The Johns Hopkins University Press. First published in 1982.

Dugatkin, L. A. 1988. Do guppies play Tit for Tat during predator inspection visits? *Behav. Ecol. Sociobiol.,* 23: 395–399.

———1991. Dynamics of the Tit for Tat strategy during predator inspection in the guppy *(Poecilia reticulata). Behav. Ecol. Sociobiol.,* 29: 127–132.

Dugatkin, L. A., and M. Alfieri. 1991. Guppies and the Tit for Tat strategy: Preference based on past interaction. *Behav. Ecol. Sociobiol.,* 28: 243–246.

———1991. Tit-for-Tat in guppies *(Poecilia reticulata):* The relative nature of cooperation and defection during predator inspection. *Evol. Ecol.,* 5: 300–309.

Griffin, D. R. 1992. *Animal Minds.* Chicago: University of Chicago Press.

Krebs, J. R., and N. B. Davies. 1993. *An Introduction to Behavioural Ecology,* 3rd ed. Oxford: Blackwell Scientific Publications.

Maynard Smith, J. 1976. Evolution and the theory of games. *Amer. Sci.,* 64: 41–45.

Milinski, M. 1987. Tit for tat in sticklebacks and the evolution of cooperation. *Nature,* 327: 15–17.

Teale, E. W., ed. 1977. *The Insect World of J. Henri Fabre.* Boston: Beacon Press.

9. The Fine Balance between Cooperation and Conflict

Emlen, S. T., and P. H. Wrege. 1992. Parent-offspring conflict and the recruitment of helpers among bee-eaters. *Nature,* 356: 331–333.

Gadagkar, R. 1992. Disease and social evolution. *Curr. Sci.,* 63: 285–286.

———1992. When fathers harass their sons. *Down to Earth,* 1(8): 47–48.

———1995. Observational study of animal behaviour: From instinct to intelligence. *Curr. Sci.,* 68: 185–196.

Gadagkar, R., and H. S. Arathi. 1995. Complex domestic conflicts in a bird family. *Curr. Sci.,* 68: 676–677.

Gadagkar, R., and N. V. Joshi. 1985. Colony fission in a social wasp. *Curr. Sci.,* 54: 57–62.

Keller, L., and P. Nonacs 1993. The role of queen pheromones in social insects: Queen control or queen signal? *Anim. Behav.,* 45: 787–794.

Keller, L., and K. G. Ross. 1993. Phenotypic basis of reproductive success in a social insect: Genetic and social determinants. *Science,* 260: 1107–1110.

———1993. Phenotypic plasticity and "cultural transmission" of alternative social organisations in the fire ant *Solenopsis invicta. Behav. Ecol. Sociobiol.,* 33: 121–129.

König, B. 1994. Components of lifetime reproductive success in communally and solitarily nursing house mice: A laboratory study. *Behav. Ecol. Sociobiol.,* 34: 275–283.

———1994. Fitness effects of communal rearing in house mice: The role of relatedness versus familiarity. *Anim. Behav.,* 48: 1449–1457.

Moffett, M. W. 1984. An Indian ant's novel method for obtaining water. *National Geographic Research,* Winter: 146–149.

Peeters, C., J. Billen, and B. Hölldobler. 1992. Alternative dominance mechanisms regulating monogyny in the queenless ant genus *Diacamma. Naturwissenschaften,* 79: 572–573.

Peeters, C., and S. Higashi. 1989. Reproductive dominance controlled by mutilation in the queenless ant *Diacamma australe. Naturwissenschaften,* 76: 177–180.

Ratnieks, F. L. W. 1988. Reproductive harmony via mutual policing by workers in eusocial Hymenoptera. *Amer. Nat.,* 132: 217–236.

Ratnieks, F. L. W. and P. K. Visscher. 1989. Worker policing in the honeybee. *Nature,* 342: 796–797.

Ross, K. G. 1992. Strong selection on a gene that influences reproductive competition in a social insect. *Nature,* 355: 347–349.

Shykoff, J. A., and P. Schmid-Hempel. 1991. Genetic relatedness and eusociality: Parasite-mediated selection on the genetic composition of groups. *Behav. Ecol. Sociobiol.,* 28: 371–376.

———1991. Parasites and the advantage of genetic variability within social insect colonies. *Proc. Roy. Soc. London, B.,* 243: 55–58.

———1991. Parasites delay worker reproduction in bumblebees: Consequences for eusociality. *Behav. Ecol.,* 2: 242–248.

Slagsvold, T., T. Amundsen, and S. Dale. 1994. Selection by sexual conflict for evenly spaced offspring in blue tits. *Nature,* 370: 136–138.

Trivers, R. L., and H. Hare. 1976. Haplodiploidy and the evolution of the social insects. *Science*, 191: 249–263.

de Waal, F. 1989. *Chimpanzee Politics: Power and Sex among Apes.* Baltimore: The Johns Hopkins University Press. First published in 1982.

10. Some Caveats and Conclusions

Fletcher, D. J. C., and C. D. Michener. 1987. *Kin Recognition in Animals.* Chichester, Eng.: John Wiley.

Gadagkar, R. 1985. Kin recognition in social insects and other animals: A review of recent findings and a consideration of their relevance for the theory of kin selection. *Proc. Indian Acad. Sci. (Anim. Sci.)*, 94: 587–621.

Gadagkar, R., S. Bhagavan, K. Chandrashekara, and C. Vinutha. 1991. The role of larval nutrition in pre-imaginal biasing of caste in the primitively eusocial wasp *Ropalidia marginata* (Lep.) (Hymenoptera: Vespidae). *Ecol. Entomol.*, 16: 435–440.

Gadagkar, R., S. Bhagavan, R. Malpe, and C. Vinutha. 1990. On reconfirming the evidence for pre-imaginal caste bias in a primitively eusocial wasp. *Proc. Indian Acad. Sci. (Anim. Sci.)*, 99: 141–150.

Gould, J. L., and C. G. Gould. 1994. *The Animal Mind.* New York: Scientific American Library.

Withers, G. S., S. E. Fahrbach, and G. E. Robinson. 1993. Selective neuroanatomical plasticity and division of labour in the honey bee. *Nature*, 364: 238–240.

Index

....

*Page numbers in **bold** refer to figures and tables.*